Fascination Amazon River

Geographical overview of the fascinating world of the Amazon river from Nauta in Peru to the estuary

Lothar Staeck

Fascination Amazon River

Its People, Its Animals, Its Plants

Springer

Lothar Staeck
Berlin, Germany

The translation was done with the help of artificial intelligence (machine translation by the service DeepL.com). A subsequent human revision was done primarily in terms of content.

ISBN 978-3-662-64451-5 ISBN 978-3-662-64452-2 (eBook)
https://doi.org/10.1007/978-3-662-64452-2

This Springer imprint is published by the registered company Springer-Verlag GmbH, DE part of Springer Nature.
The registered company address is: Heidelberger Platz 3, 14197 Berlin, Germany

Contents

List of Figures

List of Tables

1

Introduction: What Travellers Can Really See on Their Trip to the Rainforest

The largest river on earth has fascinated me since my student days. Since the late 1970s, I have visited the region between *Nauta* in Peru and the estuary west of *Belem* in Brazil again and again—at least 30 times to date. Initially, I also travelled to the Amazon region for scientific purposes. The main focus of my research interest at that time was the *cichlids*, a tropical fish family from the group of perch relatives *(Percomorphaceae)*, which is characterized by an extremely interesting and highly developed courtship and breeding behavior. Of the approximately 3000 *cichlid species* worldwide to date, more than 500 have now been recorded in the Amazon and Rio Negro. Since the end of the 1980s, I have been leading natural history excursions into the vast landscape of the two river systems and have also been a tour guide and lecturer between *Nauta* and *Iquitos* (on the western edge of this river system) and *Belem* (about 120 river kilometres from the mouth of the Amazon into the Atlantic) on several occasions (see also the map on page II of this book).

On these journeys I had numerous encounters with a wide variety of animals as well as with members of different ethnic groups living on the banks of the rivers. My travels were also accompanied by a number of health risky adventures: there were bites from piranhas, caimans and tree snakes, attacks from swarms of bees and ant colonies, and a boat capsized on the Amazon. Fortunately, I survived all of it ultimately unscathed. To report about all this in detail would go beyond the scope of this work.

The motivation to write a summary volume on the fauna and flora of Amazonia, including the people living in this region, instead came from outside. Travel participants and listeners of my lectures have repeatedly asked for a summarizing presentation

L. Staeck, *Fascination Amazon River*, https://doi.org/10.1007/978-3-662-64452-2_1

- the fascinating flora and fauna,
- of the native peoples and the *Caboclos*, the descendants of original immigrants from Europe and these same native peoples, as well as ultimately,
- of the mighty Amazon River itself, with its enormous seasonal variations, its extremely diverse tributaries, and its general imposing characteristics

in demand. Unfortunately, this entire region is extremely endangered by constant slash and burn of the forests, by overexploitation, by illegal massive exploitation of mineral resources (e.g. gold), by the construction of huge dams, but also by the exponential growth of the human population in these sensitive ecosystems—for example, the population of Brazil has quadrupled from 1951 to 2018, from 51 million to now over 210 million. No one knows how much longer it will take until the river and its adjacent rainforests collapse. But one thing is certain: if the listed facts for the forest destruction and the river pollution and above all the exorbitant population increase do not stop by the time the next generation grows up, then in the next century there will be no more primary forests on the Amazon, and with them almost all wild animals will then also be extinct.

Perhaps the present volume will help to avert the great catastrophe that is looming in Amazonia.

I have been traveling in the Amazon lowlands for many, many years. Every time I came back, I thought on a newly started trip that I had now seen everything and there was nothing new for me to experience. But then on each subsequent trip I was surprised anew that—in addition to meeting foreign people—I kept encountering flowering plants and animals that I had never seen with my own eyes before. That's why every trip to the Amazon is full of surprises!

From the immense wealth of plant and animal species that are still far from being comprehensively identified, this book filters out, in characteristic photographs and exciting descriptions, those that visitors can actually see on a 3-week journey through the area. Each of the plant and animal examples selected in this volume has an astonishing, often unbelievable peculiarity—either in its special morphology or in its behaviours, which make you want to read or think further about the featured plant or animal.

In my research into the classification of certain plants and animals, I have discovered that there is currently a huge correction or reclassification of the long established *nomenclature*, the internationally binding scientific list of names of living things. New methods in the study of phylogeny and from molecular biology make it possible to analyse the genetic material—the genome—of a particular living being in detail. In the process, numerous new relationships are currently being established and rearrangements as well as new species definitions are being made, which are also taken into account here.

However, the focus of this book is, as mentioned, on those plants and animals that the visitor can actually see during a flying visit. The large mammals such as the jaguar, the puma, the ocelot or the lowland tapir are therefore left unmentioned. One would first have to penetrate deep into the jungle and also live there for a longer time in order to have the chance to meet them at all. Nevertheless, the traveller will come across their traces more often, e.g. at markets where their skins are offered or tapir meat is on the menu, for example in *Iquitos*.

Dive now with me into the fascinating world of the Amazon from *Nauta* to the mouth (please see frontispiece).

2

The Amazon: The Largest River on Earth

The Amazon is a river of extremes. In this chapter, its statistical data are described in detail. The reader should keep in mind, however, that all these data are provisional or approximate, as the data situation for this huge area is still incomplete today. Furthermore, the course of the river is traced from its source via various tributaries to its mouth. In the course of the last 25 million years, the geological development of this river has undergone an incredible turnaround, which can still be seen in many places on its banks. In addition, a large number of the creatures found in the river today document this reversal from a marine bay to a freshwater river. Due to the seasonal fluctuations in the river level, the morphology of both sides of the river bank is constantly changing. Both on the so-called river cliff, the steep bank on the outer side of a river bend, and on the slip-off slope, the bank on the opposite inner side, new landscape forms and river courses have developed over the years (see also Sect. 2.4). In addition, typical habitats develop during the rainy season: the flood meadows *(Várzeas)* and the flood forests (*Igapós*). These and the dry high forest *(terra firme)* are home to specific fauna and flora. Finally, the Amazon river system includes rivers with completely different chemical compositions. Thus, there are white-water, black-water and clear-water rivers, which allow the existence of very different species of plants and animals.

The largest river on our planet, with its approximately 10,000 direct or indirect tributaries and adjacent rainforests with an original size of 7,500,000 km^2, constitutes the huge Amazon ecosystem directly on the equator. Figure 2.1 shows how extensive this area is: The Amazon basin including all rainforests can easily fit into the land area of the United States (without Alaska and Hawaii) with its approximately 8,000,000 km^2. There are certainly

L. Staeck, *Fascination Amazon River*, https://doi.org/10.1007/978-3-662-64452-2_2

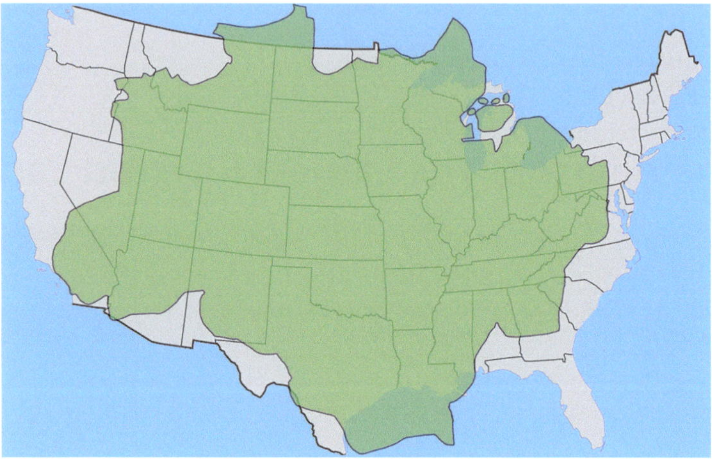

Fig. 2.1 Size comparison of Amazonia and the land area of the United States (without Alaska and Hawaii)—Amazon basin is marked in green colour

still many unknown animal and plant species in this vast area, which has still not been fully explored. To date, more than 5500 vertebrate species and more than 56,000 seed plant species have been described in Brazil alone (for comparison, in Germany: 530 vertebrate species and about 3500 seed plant species)—and every year numerous newly discovered animal and plant species are added. The Amazon basin is already ranked first in the world for plant, mammal, fish and insect species. Probably all the species of the Amazon lowlands will never be discovered.

Among the indigenous people, the Indians, even today every few years are discovered completely isolated groups living in the rainforest without contact with modern civilization. For example, newspapers recently reported that in the Brazilian Amazon, indigenous groups spotted from a helicopter shot at the explorers' small plane with bows and arrows. According to estimates by the Brazilian Agency for the Protection of Indigenous Peoples (FUNAI), there are over 100 isolated indigenous peoples living in the Amazon rainforest alone, whose language and way of life are completely unknown.

This is all possible despite our technologically advanced and connected world, because the Amazon rainforest is so incredibly vast and yet so difficult to access. In addition, the tropical climate is extremely hostile to humans.

2.1 The River Between Its Source and Mouth

Since GPS calculations by Brazilian scientists in 2012, the Amazon River, which flows continuously through the South American continent from west to east at the height of the equator, is considered the longest river in the world at 6868 km, thus displacing the Nile with its 6695 km in second place (see also Table 2.1).

The source was located in 1996 in the *McIntyre Lagoon*, 200 km west of Lake Titicaca high up at 5547 m on Mount *Nevado Mismi* in the Peruvian Andes, less than 200 km from the Pacific Ocean between the cities of *Cuzco* and *Arequipa*. From there, more and more streams and rivers join together to form a common drain that flows down from this massive mountain range. This watercourse at first, bears many different names, until at last the name Amazon is found on the map: *Lloquera* at first, then *Callamayo, Hornilles, Rio Ene, Apurímac, Rio Tambo, Ucayali* and *Marañon* near *Nauta* in Peru, where the *Ucayali* and the *Marañon* join at the small village of *Puerto Grau*, it is called Amazonas for the first time by the Spanish-speaking people (please see page II).

There are a number of interpretations of the name "Amazonas". It is possible, for example, that it is derived from the *Tupi-Guarini* language term "*Amaçu*", which means "sound of water" and refers to the tidal wave of the sea, which moves upstream from the Atlantic every year in a superposition of new moon and high tide effects from two tributaries of the Amazon. Even at high tide, the water depth at *Nauta* is often only about 6 m, so even smaller

Table 2.1 The Amazon in figures

Length[a]	6868 km (river kilometres)
Navigable	5215 km
Tributaries[b]	Approx. 10,000
Sediment transport	900 million T/year into the Atlantic Ocean
Current	0.75–5 m/s
Water volume[c]	
At high watermark	More than 300,000 m^3/s
At low tide	About 100,000 m^3/s
Gradient from the east end to the mouth	Total 180 m = 3 cm/km
Narrowest point (near Obidos)	1.5 km
Water depth	10–30 m, at Obidos deep holes of up to 120 m

[a]Nile: 6695 km, Rhine: 1320 km
[b]Of which 15 are larger than the Rhine
[c] In the Rhine delta 2330 km/s (max. ever measured 12,000 km/s in 1926)

cruise ships rarely venture from *Iquitos* to *Nauta*. Yet this region is highly interesting, especially from a faunal point of view. For example, within the entire Amazon lowlands, some species of the spectacularly coloured poison dart frogs (tree climbing frogs) are only found in this region, from the upper reaches to the slopes of the eastern Andes (Sect. 9.1). The rare *Gray's bald-faced saki monkey (Pithecia irrorata)* is also common here.

But the Amazon did not always drain in a west-east direction to the Atlantic. Before the parts of the former primeval continent *Gondwana* drifted apart, the primeval Amazon originally flowed in exactly the opposite direction from east to west and flowed into the Pacific, because about 130 million years ago there were neither the Andes nor the Atlantic. At that time, the source of the proto-Amazon was in present-day Chad (Africa) in the *Ounianga Kébir* lakes of the *Ennedi Mountains* in the middle of the Sahara (Fig. 2.2). From its source to its mouth, this river was then the longest and most powerful river on Earth at 14,000 km. With the break-up and drifting apart of the parts of the primeval continent, the Atlantic Ocean was formed and in the west of the new South American continent—beginning about 25 million years ago—the Andes gradually folded up, cutting off the estuary of the primeval Amazon from the sea and leaving many animal species of the sea in the now dammed up huge inland lake. In the course of the following many millions of years the water of this inland lake sweetened more and more, numerous former animal

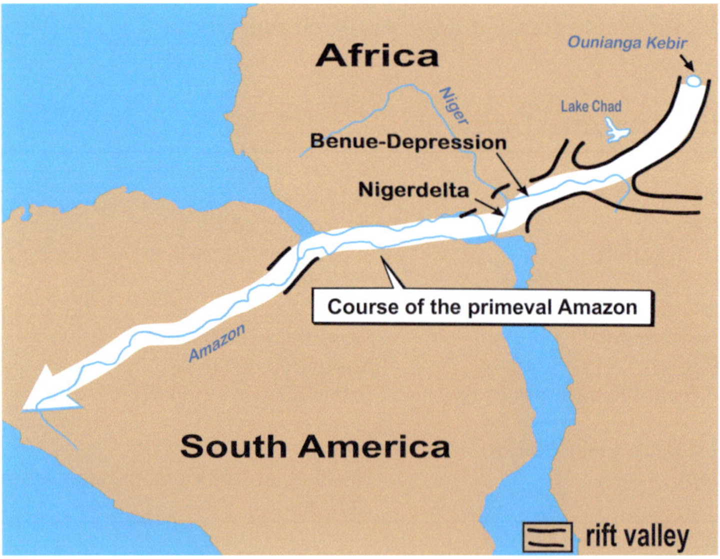

Fig. 2.2 Course of the primeval Amazon about 130 million years ago

species of the sea adapted to these changed living conditions and live in the Amazon until today, for example two dolphin species, sardines, herrings, needlefish *(Belonidae)*, boxfishes, rays, shrimps and crabs in more than 30 different species, sponges and many other former inhabitants of the seas. The pressure of the water masses collecting in the inland lake finally became so strong that the water now drained eastward in the same glacial valley of the primeval Amazon and lastly emptied into the Atlantic Ocean. During the two ice ages in the course of the last 100,000 years, the level of the seas sank so much that the "new" Amazon plunged over gigantic waterfalls into the Atlantic at its mouth.

There is a great deal of geological and biological evidence for the existence of the primeval Amazon and thus for the river reversal, for example:

- Rock deposits from Africa in what is now the Amazon,
- The existing anomaly that the Amazon riverbed becomes narrower towards its mouth at *Belém* (and not wider, as is usually the case with large rivers),
- Finds of various fossils, for example identical species of diatoms and *mesosaurs* in the Amazon and in the desert of Chad,
- The only desert crocodiles still living today in the lakes of the *Ennedi Mountains* (Chad), genetically related to the Amazon caimans.

By the way: even today, Sahara dust fertilizes the rainforests of the Amazon basin, as recent meteorological studies have shown. With the constant northeast trade wind, the incredible mass of 400–700 tons of it reaches every year the Amazon at an altitude of 5 km. Without this gigantic fertilizer load, the vegetation on the extremely nutrient-poor soils would never grow as lushly as we find it today.

Today, the Amazon lies in a vast lowland plain covered with sediments and sedimentary rocks, bordering the Andes to the west, the ancient low mountain ranges of the *Guyana* lands to the north, and the Brazilian Shield to the south, an ancient, eroded mountainous land. In some sections of today's Amazon, one can still see the former steep banks of the primeval Amazon: huge, million-year-old laterite cliffs rising up to 15 m, in which Amazon *Ringed kingfishers* today dig their breeding burrows (Fig. 2.3).

The slope of the river from the foot of the East Andes to its mouth is only an incredible 180 m, which is only 3 cm/km of river! Despite this extremely low gradient, the river flows at 0.75 m/s even in the dry season, which is 45 m/min or 2.7 km/h. In the rainy season, when the water masses from the many tributaries push into the main stream, the flow velocity increases up to 18 km/h. Due to the ongoing deforestation of the rainforest, the forest loses

Fig. 2.3 Remains of a laterite cliff of the proto-Amazon with breeding burrows of the *Ringed kingfisher* (Sect. 8.5.6)

more and more of its ability to store rainwater every year, so that the rainfall reaches the Amazon faster and faster, where it increases its flow velocity. The highest flow velocity recorded so far was 39.2 km/h. The reason for such high speeds are the extremely large amounts of water that flow "down" the river even in the dry season. At low water it is around 100,000 m^3/s, which can swell to more than 300,000 m^3/s at the peak of the rainy season. In comparison, the amounts of water that, for example, rush down the Rhine Falls near Schaffhausen in Switzerland during the summer are hardly worth mentioning at 600 m^3/s. The highest flow ever recorded there was in the summer of 1965: 1250 m^3/s. During the rainy season, grass-reed islands of various sizes with numerous common water hyacinths *(Eichhornia crassipes)* can be seen constantly floating downstream on the Amazon, torn loose from the bank (Fig. 2.4; Sect. 6.4.2). Some of these floating plant islands may well be up to 100 m long. Once I experienced a loud concert of different "musicians" on such a drifting island: there were the bright voices of small frogs mixed with the deeper voices of toads, accompanied by the chirping of male cicadas. Already from a distance I heard this many-voiced chorus, which swelled to a loud roar when the island swam directly past, and then finally gradually became quieter and quieter.

About 120 river km further east of *Nauta* near *Iquitos* (Peru), the width of the Amazon is already 1.8 km. Near the metropolis of *Manaus* in Brazil, the black-water *Rio Negro* flows into the Amazon, which is now 5 km wide. For

Fig. 2.4 Grass-reed islands drifting downstream during the rainy season

the Brazilians the river is called "Amazonas" only from *Manaus* on, from *Iquitos* to *Manaus* they call it *Rio Solimões*. *Manaus* is only about 20 m above sea level, but the linear distance from this city to the mouth of the Amazon into the Atlantic is still 1370 km. Due to the large amounts of rain of up to 3 m/year, which fall mainly between February and June, the river level in *Manaus* rises by up to 15 m at the end of the flood season, about the end of mid-June. The flow velocity of the water draining to the Atlantic is therefore up to 18 km/h at high tide east of *Manaus*—despite the low river gradient.

Near *Belém* at the beginning of the Amazon estuary,which is correctly an estuary of the *Rio Pará*, extends the Bay of *Guajará (Bahia do Guajará)*, which is a part of the *Rio Pará*. This is a large southern branch of the Amazon, which has joined with the *Rio Tocantins* and some smaller rivers. Nowadays, hardly any water flows from the Amazon into the *Rio Pará*, at most via some of the narrow *Brevis channels*. Thousands of years ago the southern arm of the Amazon estuary was blocked by the mud and sediment masses flowing in. The bay of *Guajará* is 4 km wide at the level of the port of *Belém*. On the other side of this bay is an island, behind which flows the *Rio Pará*, which is about 15 km wide there. Afterwards, the *Rio Pará* becomes wider and wider until it is up to 80 km wide in the estuary. In this estuary lies also the island *Marajó*, which is 40,100 km^2 as large as Switzerland. It is thus bounded to the east by the *Rio Pará* and to the northwest by several narrow and also very wide estuary arms of the Amazon, which currently represent the actual outflow of the Amazon. South of it flows a series of very narrow channels, often only 100 m wide, called *Brevis channels*. These scenically extraordinarily attractive natural channels, where the virgin forest reaches the banks—which means that it is very close from the ship—connect *Belém* and the Amazon to the west. If you add up all these outlets of the Amazon's funnel mouth to the west and east, you get a total width of up to 250 km. Still more than 200 km to the east and north to French Guyana, the Atlantic is coloured clay-yellow by the sediments of the inflowing Amazon, which add up to 1.2 billion tons per year.

Conversely, the tides make themselves be noticed surprisingly far inland. While the average tidal range, i.e. the difference between high and low tide, at *Belém*—about 100 km linear distance from the mouth of the Atlantic—is 3 m, at *Santarém* (750 km linear distance from the Atlantic, please see page II) it is still 0.35 m. Even in *Obidos*, 800 km linear distance from the Atlantic, the tides are still noticeable at 0.10 m. For comparison: in Hamburg-Altona, Germany, the average tidal range is 3.60 m at a distance of 92 km from the North Sea.

The river's inhabitants also affectionately call the Amazon "*El Rio Mar*", the ocean river, because its length and width as well as its water masses truly resemble a sea or estuary rather than a stream. When travelling on the Amazon by boat, the visitor- rarely sees—or glimpses—both banks of the river. There are almost always many elongated islands in the river, most of which block a view of the opposite bank. Thus, the actual width of the Amazon is rarely visible to the eyes. In addition, during the rainy season the water surface doubles from 300,000 km^2 to 600,000 km^2. For comparison: The total area of the US-state of Arizona is 295,000 km^2. Due to these immense water masses, even the earth's crust lowers by 8 cm during the rainy season—as recent GPS calculations have shown.

2.2 Várzeas: Floodplain Meadows

The river regularly overflows its mostly shallow banks between April and July. In these months the water masses of the main stream pour out on both sides of the river up to 50 km into the flat land, where in the dry season wide meadow areas with water-poor, natural channels extend. These countless channels—known as *igarapés* (Brazilian for a narrow, natural channel)—which connect the tributaries and lakes to each other, then fill with water.

After long periods of heavy rainfall, smaller tributaries and even these *igarapés* temporarily reverse their direction of flow. This creates a huge lake landscape of floodplain meadows—so-called *várzeas* (= Portuguese for river floodplains)—covering an area of 200,000 km^2 (Fig. 2.5). This vast open lake landscape is characterized by a high nutrient content. The sediment-laden water masses flowing in from the main stream literally fertilise these flooded meadows, which are only a few metres deep, with the result that an extremely nutritious food web of up to 15 different species of floating leaf plants (Sect. 6.4), dozens of fish species (Sect. 8.2) and just as many predator species—especially birds (Sect. 8.5)—can develop.

At the shallower banks of the normally wall-like embankment, the water masses of the main stream constantly flow at great speed into the hinterland as the rainy season progresses, where they form to the *várzea* ecosystem (Fig. 2.6). With the water flow, many fish enter these *várzeas*, where they spawn in the calm shallow water zones among the aquatic plants. The largest water lily in the world, the giant Queen Victoria's water lily *(Victoria amazonica)*, also grows in these current-free *várzeas* (Sect. 6.4.1). On elevations within these floodplain meadows, as well as on riparian slopes, trees grow primarily as pioneer plants because, as fast-growing, sun-loving plants, they are the first to establish themselves on land that becomes available. Characteristic pioneer trees of the *várzeas* are the slender "ant trees"—other names: snakewood, guaramo—(*Cecropia*) with their light-coloured trunks

Fig. 2.5 *Várzea lakescape* of floodplain meadows; the arrow marks the border of the riverbed

Fig. 2.6 Huge masses of water flow into the flat *hinterland*, where they form the *várzeas*. In the background on the right, numerous giant water lilies *(Victoria amazonica)* grow

and chestnut-like leaves (Sect. 6.1.1 and Fig. 6.1), but also *Pseudobombax munguba*, *Ocotea pulchella* as well as *Campsiandra comosa*, Humboldt's willows *(Salix humboldtina)*, *balsa trees (Ocrhroma pyramidale)*, the Spanish cedar or cedro *(Cedrela odorata)*, various *Ficus* and *Inga* species (Chap. 6). Thus, in some sections of the *Várzea*, young forests with these pioneer woody plants also develop over the years, which are also submerged during the flood season. At the peak of the flood, the water level remains for up to 3 months. As the dry season progresses, starting around the end of August/September, the *várzeas* gradually run dry again, and the adult fish and the many young fish must increasingly hurry to get back into the open water. Those who do not succeed in doing so, or who do not have the organs to survive the dry season, now become easy prey for flocks of herons, cormorants, great-billed terns and other bird species that fish the remaining pools.

2.3 Igapós: Floodplain Forests

On other sections of the riverbank, however, the rainforest, which has grown over centuries, reaches the banks of the main river with a few giant trees. When the river then overflows its banks with increasing high water and pours into the hinterland, even this old rainforest is flooded for kilometres during 7 or 8 months, so that the trees are thus submerged for a long time (Fig. 2.7), lower trees even up to the crown. The trees do not mind these long-lasting

Fig. 2.7 The flooded forests of the Blackwater rivers and lakes (Blackwater Igapós) are under water for up to 7/8 months

floods. In Central Europe, only white willows *(Salix alba)* can withstand such long periods, while cottonwood *(Populus)* begin to rot after about 4 months, English oaks *(Quercus robur)* and elms *(Ulmus minor)* after just under 3 months and sycamores *(Acer pseudoplatanus)* after only 15 days.

This special type of landscape in Amazonia is called *igapó* (= "forest in the water"; *Tupi* language). In the area of the blackwater rivers and lakes the *igapós* look particularly dark and almost mysterious (Fig. 2.7). These seasonally flooded trees are mostly lower, more poorly developed, and often gnarled in growth, and the number of tree species in this riverine landscape is also much lower than in the never-flooded rainforests of the *terra firme*. More on this later.

The submerged trees are capable of anaerobic metabolism during this period, which means that they are not dependent on a direct oxygen supply. On the other hand—as described—their growth is greatly reduced during this period. In addition, the fine root system of particularly adapted tree species increases considerably (surface enlargement), as the fine roots sprout directly from the trunk like large beards under water and absorb the oxygen dissolved in the water. When the water recedes again and the ground dries out, the roots that have been preserved during the rainy season immediately begin to grow up to 15 cm in depth.

During the flood season, the monkeys and birds have to share these trees with the fish. Because of this periodic flooding of the rainforest, all its inhabitants can also swim, even sloths and monkeys, as well as insects, for example ants. Many ground dwellers of this forest have even acquired strategies against flooding in the course of evolution, for example:

- Some can temporarily walk on the water surface with the help of an air film in their body hair (such as isopods, mites, ants);
- There are insects in which only 10% survive the flooding of their habitat, but the survivors produce so many already developed offspring immediately after the water recedes that the same population density is reached as before the flooding comes;
- Conversely, siliceous sponges, for example, survive the dry season in permanent cysts (strangulations) on tree trunks. With the onset of the flood tide, they awaken to active life again.

During the rainy season, the flooded forests result in an additional water surface of 100,000 km^2. Thus, with increasing duration of the rainy season, the water masses of originally 300,000 km^2 plus 200,000 km^2 *várzeas* plus 100,000 km^2 *igapós* add up to a total of 600,000 km^2 (= twice the total area of the US state Arizona). Thus, for 7 months during the rainy season, very different landscapes are created on both sides of the main river.

2.4 River Meanders and Terra Firme

Due to the low gradient of the Amazon, it forms many meanders as it flows towards the Atlantic. On the bank side, where the current with its enormous water masses hits the bank most directly during floods—the outer radius of the river—a baffle bank (or baffle slope) is formed (Fig. 2.8). This is the "death

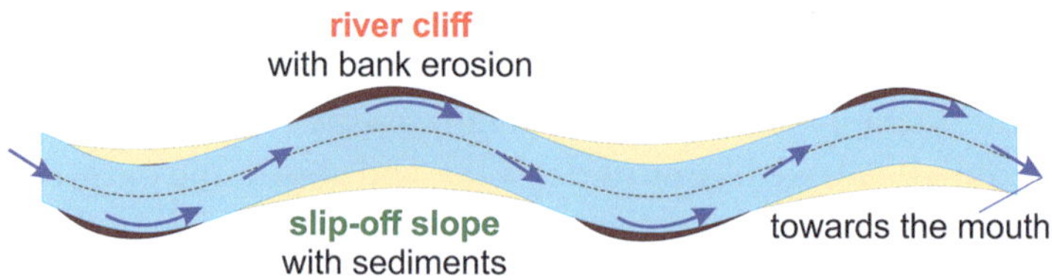

Fig. 2.8 Schematic representation of river cliffs and slip-off slopes: as the rainy season progresses, the river cliffs are eroded and bank failures occur. During the dry season, sediments are deposited on the slip-off slopes, so that sandbanks can gradually form

zone" of the rainforest, because here violent bank erosion of the soft, clumped sediment base occurs as floodwaters increase. As a result, many trees standing at the edge of the bank are uprooted every year and fall into the river when the watcr runs off (Fig. 2.9).

The river thus becomes wider and wider on the river cliff. So it can happen that the river meanders come closer and closer to each other. Then one day the water masses of the main stream break through the bank and look for the shortest way through the rainforest following the gradient. In this process, neighbouring river meanders meet or merge. The river flows from now on through this shortcut. Thus it can happen that pieces of forest on one side of the river suddenly find themselves on the other side. New small-scale ecosystems are created, where now biological different species meet and possibly cross, which never met before because of the river as a natural boundary. On the other hand, members of the same species are also permanently separated from each other by this event and can thus become new species over time after many generations.

In the course of the following years, the river loop, which is no longer traversed by the current, develops into a gradually silting up and overgrown almost circular oxbow of the river (Fig. 2.10). This can be easily observed in many places, especially from an airplane.

Fig. 2.9 In the "death zone" of the river cliff, many trees fall into the river every year

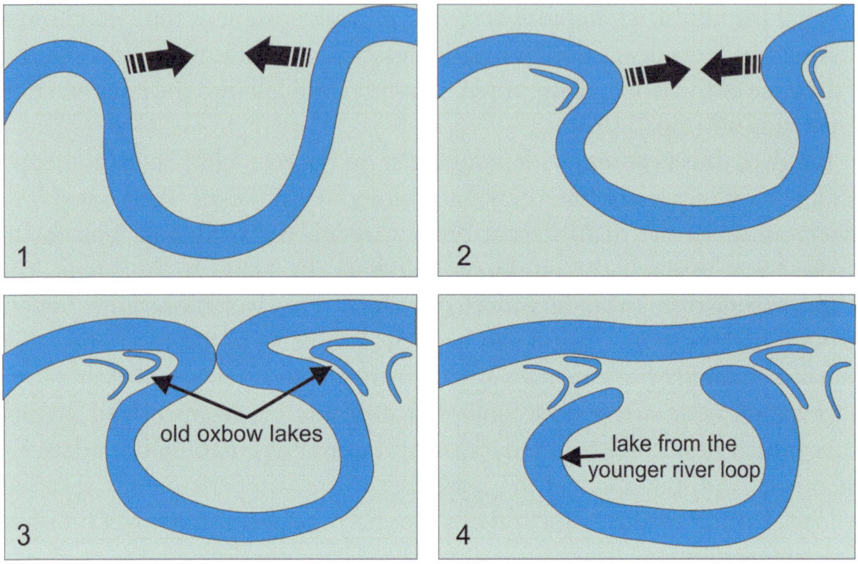

Fig. 2.10 Schematic representation of the water breakthrough at a river bend

On the low-flow bank edge of the opposite side (the inner curve bank) the water masses flow quietly along the bank without eroding it. This bank on the inner radius of the Amazon is called the slip-off slope. Here, in contrast to the river cliff, sediments are constantly accumulated, so that large sandbanks can form in the dry season. In these places, the establishment of pioneer species (such as *Cecropia*) can be observed very impressively. These trees win the first round of colonization; however, shade-tolerant species gradually grow up under their canopy, replacing the short-lived pioneer species after 10–20 years.

During the dry season between October and December, steep banks of up to 15 m in height develop at the edges of the main stream, so that people have to dig many steps into the soft sedimentary soil or build ladders to reach their gardens or houses during this period (Fig. 2.11). In contrast, when the floods approach their peak in May and June, the river water flows over the edge of the bank into the land behind it, which lies somewhat lower, flooding banana and cassava plantations or cattle pastures as well as the villages near the banks. Although people build their houses on stilts for this reason, many people have to leave their homes in the event of extreme flooding, which occurred with increasing frequency in recent years. The landscape along the Amazon is therefore totally different during the dry and rainy seasons (Figs. 2.12, 2.13 and 5.7).

Fig. 2.11 During the dry season, a steep bank rises up to 15 m high at the river edge

Fig. 2.12 Bank edge at low water

Fig. 2.13 Riparian edge during floods

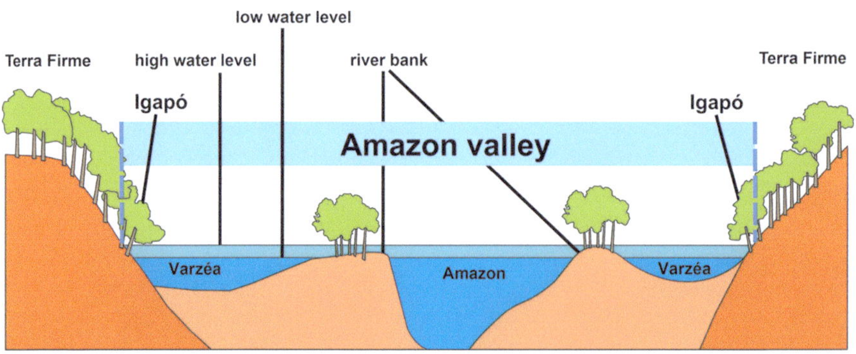

Fig. 2.14 Cross-section of the river landscape with its ecosystems on both sides of the Amazon River

The relief of the Amazon with its different landscapes during the dry and rainy seasons is shown in Fig. 2.14.

Outside the floodplains the actual rainforest extends. This ecosystem lies so high that it is never flooded. Therefore, this type of landscape is called *terra-firme* (= solid land). It is not only significantly more species-rich than the *igapó forests*, but these rainforests are centuries old and therefore the most species-rich forests on our planet.

The Amazon is navigable all year round between Manaus and its mouth. However, there is little shipping traffic. It is limited mainly to a few passenger ships and regional cargo ships that connect the settlements and towns along the river, as well as to a total of a few container ships and cargo ships per day that bring goods mainly to and from Manaus—for example tropical timber, raw materials such as bauxite, iron ore, hard coal, gas or oil. In the rainy season, ocean-going vessels of up to 9000 gross registered tons can even sail to *Iquitos*, which is 4200 river kilometres from the Atlantic. Iquitos is thus a seaport in the middle of the continent. In addition, the river is still navigated by the typical multi-storey wooden Amazon ships with an oval superstructure, mostly belonging to *caboclos* (see also Sect. 4.2), which are used for transporting smaller loads and sometimes also as livestock transporters (Figs. 2.15 and 2.16). Finally, dugout canoes or canoes with outboard motors are occasionally seen as a means of transport between villages. During the rainy season, some cruise ships of different shipping companies also sail mostly as far as *Manaus*, more rarely as far as *Iquitos*. Except for the large container and cruise ships, the boats always sail close to the shore, where the water resistance is lower upstream than in the middle of the river.

Animals are few on the main stream—except for the large-billed terns, because the current is too strong for them. In the course of a trip you can see at most once in a while a pink river dolphin *(Inia geoffrensis)* or a giant otter *(Pteronura brasiliensis)* on the shore, scarlet macaws *(Ara macao)* and different Amazon parrots fly over the river.

Fig. 2.15 Cattle transport on the Amazon River

Fig. 2.16 Typical multi-storey Amazon ship

2.5 White Water

According to the characteristic degree of transparency and inherent coloration of the river water three different types of water bodies are distinguished in the Amazon region. The differences for these different water colours are based on geological, landscape and climatic characteristics (Fig. 2.17). The Amazon is a white-water river, that means its water colour is clayish-yellow due to many inorganic suspended solids, mostly sediments, which are washed into the Amazon tributaries from the Andes (Fig. 2.18). In one litre of Amazon water are floating about 3 g sediments, respectively in one pound Amazon water are floating about 1.5 g of sediments (glacial milk, sand and clay particles). Visibility is low and amounts to a maximum of 0.50 m. Up to a depth of 10 cm only 44% of the sunlight reaches the surface, up to 20 cm only 20%. This is why there are no submerged aquatic plants in white-water rivers, but numerous floating aquatic plants due to the many nutrients. In addition water snails are rarely found in white water and plankton not at all. Chemically, white water is almost neutral (pH = 6.5–6.9). Further chemical characteristics can be found in Table 2.2. Other Amazonian white water rivers are, for example, the *Rio Madeira* as well as the *Rio Branco* and almost all its headwaters in the Andes. In Germany, for example, the Inn and Danube are white-water rivers in sections, and in Switzerland the headwater of the Rhine.

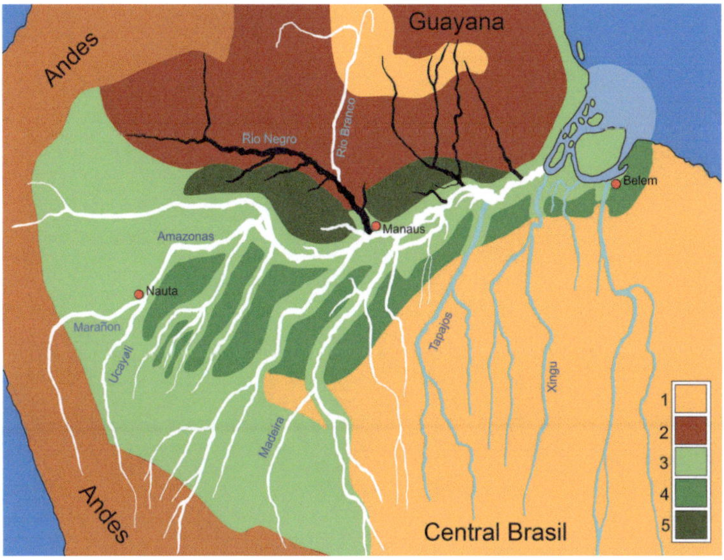

Fig. 2.17 Distribution of the three types of water bodies in the Amazon basin: 1 = Ancient degraded uplands (Guyanas and central Brazil), 2 = Swampy lowlands, floodplains and pale sands, 3 = Floodplain forests (white-water Igapós*) and terra-firme on alluvial sediments of the Andes, 4 = Rainforests (terra-firme), 5 = Floodplain forests (mostly black-water Igapós)*

2.6 Black Water

A considerable number of blackwater rivers and *igarapés*, as well as outflows from blackwater lakes flow into the Amazon. The water colour of the clear, transparent water is not really black, but rather tea- or cola coloured, sometimes also reddish, with a visibility under water of up to 2 m (Fig. 2.19a). Vast quantities of decomposed plant material are responsible for this colouration. These deceased plant fragments get to the water after the floodplains were inundated, where they form tannins as well as humic and fulvic acids. Sediments—as in the geologically young Andes—do not exist in the blackwater regions. So it happens that near the Amazon in the middle of a *várzea* or even in the rainforest are blackwater lakes or small blackwater rivers. These are formed when, year after year during the rainy season, large quantities of leaves and other plant material from the densely vegetated surroundings are washed into these waters. These gradually become more acidic and "black" (Fig. 2.20). Chemically, blackwater is extremely acidic, with a pH of 3.8–4.9, and thus hostile to many organisms, for example, the eggs and larval stages of mosquitoes. Trees living in the environmental area of blackwater are also usually

Fig. 2.18 With its typical clay-yellow coloration, the Amazon is a white-water river with its typical clay-yellow coloration

significantly lower and weaker structured than those living near whitewater. Further chemical characteristics can be found in Table 2.2. Other blackwater rivers in the Amazon are the *Rio Negro*, the *Rio Jutaí* and many other rivers originating from the swampy lowlands and bleaching earth areas north of the Amazon. In Germany, for example, the river Ilz near Passau is a blackwater river. Bogs and swamps also have often blackwater.

2.7 Clear Water

Finally, there is a third type of water body in Amazonia: the clear-water rivers. These are light green to yellow-green in colour due to algae growth on the bottom and are of exceptional clarity because there is almost no sediment or dissolved organic matter (Fig. 2.19b). Visibility is more than 4 m. The headwaters of these rivers are the ancient, eroded mountain massifs of the *Guyanas* in the north of the Amazon or central Brazil in the south of the Amazon, where the sediments and eroded sand masses have long been removed. Chemically, the water is slightly acidic (pH = 6.4–6.7). For other chemical characteristics, see Table 2.2. Important clear water rivers in the Amazon are the *Rio Tapajos* and *Rio Xingu*, which flow into the Amazon from the south.

Table 2.2 Geological, chemical and landscape comparison of the three Amazonian water body types

	White water *Agua branca*	Black water *Agua preta*	Clear water *Agua clara*
Visibility	10–50 cm	1.5–2.5 m	More than 4 m
Watercolor	clayish yellow, murky	Tea- or cola-coloured, sometimes reddish	Greenish
Source area of the river	Andes	Floodplains, marshlands	Mountainous region of the Guyanas, central Brazilian massif
Soil structure	Latisols, clay minerals, clays, end moraines	Bleaching sand, podsols	Latosols, clays, brown clays
pH value	6.5–6.9	3.8–4.9	6.4–6.7
Total hardness (in dH)	0.65–1.27[a]	0.09–0.45[a]	0.31–0.82[a]
Conductivity (in μS/cm³)	30–70[b]	8–15[b]	15–30[b]
Shoreline	Lush, *várzeas*, *igapós*, also giant trees	"*Caatinga*"-like, low trees, also shrubs, white beaches	Low trees, shrubs
Aquatic plants	Many floating aquatic plants in oxbow lakes and *várzeas; Victoria amazonica*	Few floating aquatic plants, e.g. bladderworts *(Utricularia)*	None

[a]Content of calcium and magnesium; in Germany 12–18°dH
[b]Total dissolved minerals; in Germany 300–500 μS/cm³; distilled water: 0 μS/cm³

In Germany, for example, the river Isar and in Switzerland the headwaters of the Rhine, are clear water rivers.

Of course, these three different types of water bodies have far-reaching consequences for the entire plant and animal life in Amazonia. There are also all kinds of mixed forms between these water bodies, for example when different types of water flow together.

Fig. 2.19 (**a**) A typical blackwater river with the characteristic cola-like colouring, the tannins sometimes staining the substrate reddish. (**b**) The *Rio Tapajos*, a typical clear-water river, which appears light green to yellow-green due to algae growth on the substrate

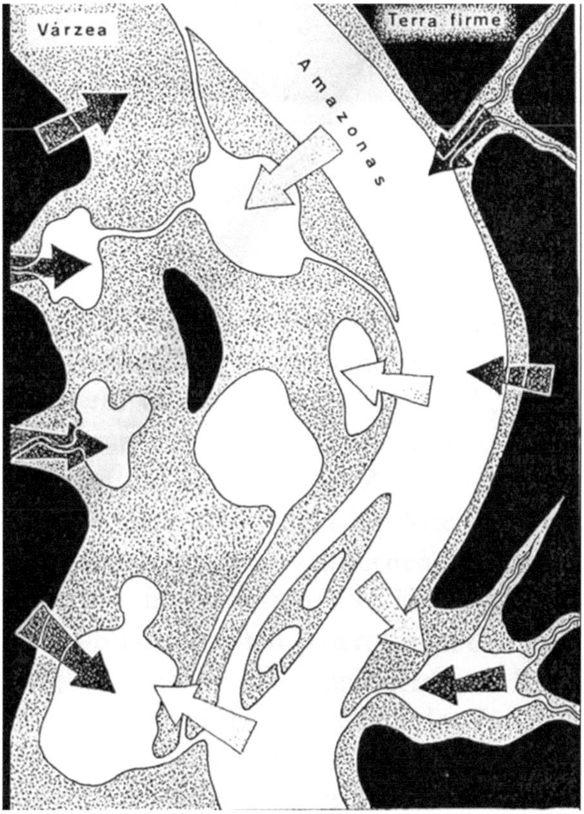

Fig. 2.20 Schematic representation of acidification in lakes near the Amazon

3

Cities Along the Amazon River

All settlements and cities along the Amazon have no road connections to the interior country. Nevertheless, there are cars and especially motorcycles everywhere today, which make life easier for the people in this hot and humid climate. All goods traffic on the Amazon, on the other hand, is not carried out by road but by ships, and to a lesser extent by air or via grass tracks. This chapter describes the lives of people in the following cities: *Iquitos* (Peru), *Leticia* (Colombia), *Manaus* (Brazil), *Parintins* (Brazil), *Santarém* (Brazil) and *Belém* (Brazil). Above all, the markets provide a good insight into the reality of the life. In addition, interesting sights and attractive excursion destinations are presented.

3.1 Iquitos

Iquitos (Peru) is only 92 m above sea level, although the mouth of the Amazon is almost 3000 km linear distance away (4200 river kilometres). The Amazon, which flows not far from the port, already has a width of 1.8 km here in the dry season and over 7 km at the peak of the rainy season in June. Due to the gushing oil production and the booming tourism, the city has experienced a huge population growth and today—situated in the middle of the (now mainly secondary) jungle—has more than 400,000 inhabitants. Car traffic has also increased extremely in recent years within the city, mainly due to the small *motokars*, three-wheeled taxis, of which there are more than 40,000 (Fig. 3.1), although there are no road connections through the jungle to other regions.

L. Staeck, *Fascination Amazon River*, https://doi.org/10.1007/978-3-662-64452-2_3

Fig. 3.1 More than 40,000 three-wheeled *motokars* drive through the streets of Iquitos

The city can be reached via the international airport. There are daily connections to Miami and Lima. On the other hand, the port is connected to a large number of cities and settlements via numerous ferry connections; there are also speed boats that travel 50–70 km/h fast—even at night for a hefty surcharge.

Iquitos is considered the gateway to the Amazon. The markets are overflowing with products from the forests and rivers such as wood, fruits, for example bananas, pineapples, papayas, avocados, guavas, Brazil nuts, cashew nuts, passion fruit or açai berries, vegetables such as palm hearts, sweet potatoes, manioc, taro, tucumão and pupinha (Sects. 7.3.1 and 7.3.2), prohibited skins, such as those of the ocelot, the anaconda and the black caiman (jacaré), as well as fish such as arapaima, tucunaré and tambaqui (Sect. 8.2), and imported goods from China, such as clothing and household items. The menus of the small snack bars and restaurants also include unusual offerings such as tapir meat, caiman steak, snake stew, turtle meat and turtle eggs and fat *suri* grubs (Fig. 3.2; Sect. 6.1.13), the larvae of the palm weevil.

The city has a number of interesting sights to offer. For example, in the center *Plaza de Armas* is situated. At this place visitors can see trees covered with interesting epiphytes as well as the imposing *Iglesia Matriz* (Mother

Fig. 3.2 The fat *suri* grubs of the palm weevil are considered a delicacy by the natives of Iquitos

Church). At its eastern corner a worth seeing building is situated, completely made of cast iron: the iron house called *Casa de Fierro*, designed by the French architect *Gustave Eiffel*, the builder of the Eiffel Tower in Paris (France). A block away from this plaza to the east is *Malecon Tarapaca*, a waterfront small boulevard that was redesigned a few years ago. At this road visitors can admire a magnificent building (the former Hotel Palace) built erected in the Art—Nouveau—style during the rubber boom. Its facade is covered with pretty tiles from Portugal. On the *Malecon* you find also the renovated *Café Fitzcarraldo*, where Werner Herzog, Klaus Kinski, Claudia Cardinale and the rest of the film crew met during the shooting of the film "Fitzcarraldo", which is still famous in German-speaking countries and premiered in 1982. In this film Kinski plays an eccentric who wants to build an opera house in the jungle, attempting seemingly impossible things, such as pulling a typical Amazon ship over a mountain …

In the evening, young families and the youth meet on the *Malecon*. A relaxed atmosphere invites you to linger, especially since numerous bars and restaurants are open and you can enjoy appetizing exotic dishes and the tasty Peruvian beer in the open air at pleasant temperatures.

During the day you have a beautiful view down to a flood plain from the *Malecon*. In addition, you see from here also the quarter *Belén*, in which the poorer part of the population lives in mostly with palm leaves covered huts.

They stand close together on stilts in a multitude of canals and small waterways. All these houses are connected by often adventurously designed and not infrequently rotten footbridges. However, visitors should only venture into this area as well as the neighbouring huge market during the day, as pickpockets and drug addicts should not be underestimated. The market is, next to the ones in *Manaus* and *Belém*, the biggest and most diverse of the entire Amazon lowlands. All conceivable products of the tropical rainforest are offered here alongside goods for daily use "Made in China" in a fascinating ambience.

3.2 Leticia

Another 3 days downstream by cargo-passenger ship (alternatively: 10 h by speedboat) travellers will reach *Leticia* (Colombia) on the left bank of the Amazon. The distance between *Iquitos* and *Leticia* is 490 river kilometres (372 km linear distance). This "Wild West"—town is situated 82 m above sea level. The Amazon is almost 3 km wide here, although it still has more than 4000 river kilometres to go before it flows into the Atlantic.

Originally, this Colombian area near the Amazon was Peruvian, and *Leticia* was formerly called *San Antonio*. However, after a conflict with Colombia in the early twentieth century, Peru seceded this region in 1922, so that the access to the Amazon became Colombian. Even a brief war between the two countries in over this region from 1932 to 1933 did not change this situation until today. *Leticia*'s nickname is *Las Tres Fronteras*, as here the borders of the three states of Peru, Colombia, and Brazil meet (Fig. 3.3). On the other side

Fig. 3.3 Boundary between *Leticia* (Colombia), *Tabatinga* (Brazil) and *Santa Rosa* (Peru)

of the river the Peruvian settlement of *Santa Rosa* is situated. Actually, *Leticia*, together with the Brazilian town of *Tabatinga*, forms a double city, as the commercial streets merge from the Colombian to the Brazilian part, barely visible and without border controls. Both cities have recorded a strong growth of their population in recent decades, with an estimated 100,000 people now living in this twin city. As a border city with a free trade zone, *Leticia* has a brisk trade in foreign currency, gasoline, lumber, fruits, fish and illegal products (drugs) across the "green border". Rumor has it that some gangs even specialize in organ trafficking. Military and police are present everywhere in the city. So, it is better to stay in the hotel at night! But during the day, the city centre is considered safe.

Countless mopeds constantly circle the inner city streets, a road connection to the outside area does not exist. However, both cities have their own airport. There are flight connections from *Leticia* to *Miami* and the capital *Bogotá*, *Tabatinga* is only connected within Brazil, for example from *Manaus*.

The main street (*Calle 8*) leads off from the newly built *Malecon* waterfront road with moorings for smaller ships (Fig. 3.4). At this road numerous exchange offices are situated with favourable rates for the Brazilian currency real. Around this street there is a large market with separate sections for fish, vegetables and fruit as well as (mostly Chinese) goods for daily needs.

Fig. 3.4 The main street *Calle 8* in *Leticia*

A block further up, there is the *Galeria Arte Indigen*, where, in addition to modern tourist souvenirs, authentic arts and crafts products from various ethnic groups are offered—for example, from the *Huitotos*, the *Boras* and *Yaguas*. In this shop are also a small—but dusty—Indian museum and some aquariums with fish from the area. A block west of *Calle 8*, in a bank building, is the manageable Ethnographic Museum with exhibits of the *Huitotos* and *Boras*, as well as posters on the evolution of human culture over the centuries. *Parque Santander* is also not far from here. There you can admire the giant Queen Victoria's Amazon water lilies (*Victoria amazonica*), *cichlids*, and turtles in a pond. At the western edge of the park the large cathedral of *Leticia* is situated. Finally, the *Hotel Anakonda* is a good place to take a break after the sweaty tour. With a cool drink at the swimming pool, visitors can admire flowering plants of the region and some tame macaws.

3.3 Manaus

Another 4 days downstream by passenger ship—or alternatively 36 h by speedboat—is situated *Manaus* (Brazil) on the *Rio Negro*, about 16 km above its mouth into the *Rio Solimões*, 20 m above sea level. The distance from *Leticia* to *Manaus* is about 1600 river kilometres (linear distance 1106 km). The Amazon is 5 km, and it is still over 1800 river kilometres or 1370 km linear distance to its mouth in the Atlantic.

The mouth of the *Rio Negro* into the *Rio Solimões* (*Encontro das Aguas*) is truly spectacular. The boundary between the tea-colored black water of the *Rio Negro*—in Fig. 3.5, the water does not looking dark due to the blue sky—and the clayish-yellow white water of the *Rio Solimões* can be seen exactly (Fig. 3.5). Excursion boats sail directly to this spot and small aircraft fly over the confluence to provide a spectacular view of it. In Fig. 3.5, the Rio Negro enters the Amazon from the right (northwest). At the top of the picture, the Amazon arrives in two bends (from the west), then becomes very wide in the foreground, and then continues to flow to the lower left (east). The water of the Rio Negro flows along the lower left corner of the picture alongside to the Amazon water for a longer time. Depending on the amount of water of the two rivers, both rivers flow side by side for up to 11 km with a clearly recognizable border, until finally the larger, the white-water river, completely absorbs the smaller *Rio Negro*. The fact that the water masses mix only gradually is due to the completely different quality of the two bodies of water—similar to a cup of coffee into which cream is slowly poured: the flow velocity, the water temperature, the pH-value and the amount of dissolved minerals differ significantly (Table 2.2). With the confluence of the two rivers, the

Fig. 3.5 Confluence of the *Rio Solimões* (white-water river) and the *Rio Negro* (black-water river) near *Manaus*. From here on the river is called "Amazonas" by the Brazilians

Brazilians also change the name: from now on, the *Rio Solimões* is called the *Amazon* until it flows into the Atlantic.

Manaus is the capital of the state of Amazonas, which has a size of 1.56 million km^2. This makes it four times the size of Germany or Arizona/USA, but it has only 3.9 million inhabitants, which means that it is extremely sparsely populated, especially as more than two million people live in *Manaus*. Due to the rubber boom around 1890, the small village of *Manaus* became a rich metropolis with an electric tram (which ran until 1957) and street lamps. The rubber barons of the time were so rich that they shipped their linens to Europe for mangling—it didn't matter when the linens were back home. A number of luxurious buildings were erected during this period, which can still be seen in the historic city centre. *Manaus was* considered the "Paris of the Tropics" at that time!

By establishing a free trade zone in 1957, the city was gradually able to return to the rubber era, the population of this jungle metropolis continued to grow strongly and it is now considered the fourth richest city in Brazil. In the largest Honda motorcycle factory of the world, one motorcycle rolls off the assembly line every 8 s.

Following oil and gas discoveries in the wider vicinity of *Manaus* up to a distance of about 400 km, pipelines have been laid in the past 10 years through the jungle and through the rivers to the new huge terminal port on the banks of the *Rio Negro* east of the city centre, so that both crude oil and gas are pumped directly there. Large oil and gas transport ships now bring the "black gold" and natural gas from *Manaus* to *Rio de Janeiro* and other Brazilian Atlantic cities. *Manaus* is thus experiencing a major new economic boom.

Since 2012, there is a new landmark of the city: a 3.6 km long, modern bridge—total cost 1 billion Real—spans the *Rio Negro* in the north of the city. It is the only bridge so far in the entire Amazon-Rio Negro river system, which has a length of almost 10,000 km. Via this bridge, there is a road connection on the other bank of the *Rio Negro* to *Manacapuru* and to *Novo Airão*.

From *Manaus* itself there is only a 220 km road connection north to *Boa Vista* in the state of *Roraima*, and on to Venezuela, often interrupted by landslides and floods. The international airport offers connections to *Miami, and* from the port you can travel up and down the Amazon on the typical multi-deck Amazon ships.

The large market halls, in which fruit, vegetables, spices and goods for daily use are offered, extend directly at the harbour. The lovingly renovated fish hall, which was built in 1883 in the Art Nouveau-style according to the model of the Paris *Les Halles*. The designer of the Eiffel Tower, Gustave Eiffel, was the architect of this wrought-iron building at the time (Fig. 3.6).

Fig. 3.6 The now renovated Fish Hall, built in 1883 according to the model of *Les Halles* in Paris, was designed by Gustave Eiffel

Tip

A visit to a fish restaurant at the port or at *Punta Negra* (the same applies to *Iquitos*, *Leticia* or *Belém*, of course) should not be missed by any visitor. There you will find a wide selection of freshly caught typical Amazonian fish (Sect. 8.2), prepared in the style of the country and served with chips or crispy fried manioc flour. However, they are offered with their Brazilian name on the menu (Table 3.1):

Table 3.1 Names of Amazon fish in Brazilian

Brazilian	English	Brazilian	English
Tambaqui	Black pacu	Tucunaré	Peacock bass
Pirarucú	Arapaima	Jaraqui	Flag-tailed prochilodus
Pacú	Silver dollar	Surubím	Sorubim

During a walk through the historic old town, it seems particularly worthwhile to visit the legendary *Teatro Amazonas*, built in 1896 in the style of the Italian Renaissance, an opera house for about 700 visitors (Fig. 3.7). The most valuable building materials were imported from Europe, for example tapestries from France, marble from Carrara in Italy, iron and bronze from Lorraine in France, exquisite parquet flooring made of precious woods from the jungle, and tiles from Germany and Portugal. Even today the interior looks very noble. The magnificent *Palácio Rio Negro*, built in Portuguese style, also dates from the rubber boom. This impressive villa was built in 1910 by the German rubber baron Waldemar Schulz. Today it is a cultural centre and art museum. In the front garden are two smaller specimens of the Brazilwood tree (*Caesalpina echinata*; Fam. *Fabaceae*), which gave its name to what is now Brazil. The bark and resin were used for dye. Its reddish colour, reminiscent of the glow of fire, led to its Brazilian name "pau brasil" (*pau* = wood, *brasil* = glowing) and thus to the state name *Brasil* (pronounced *Brasíu* = Brazil). The pretty yellow flowers are somewhat reminiscent of orchids.

Suitable for cruise ships, the harbour, built by England in 1906, is anchored on floating pontoons that adjust to the strongly changing water level. Directly at the approaching bridge, a large sign hangs on the shore showing the maximum high water levels of the past 115 years (since 1904). The trend has been clearly upwards for a number of years. Are these ever new maxima natural fluctuations or already effects of climate change? The fluctuation between low water (in November/December) and high water (end of May/beginning of June) are gigantic and amount to up to 15 m, with the level remaining at its maximum for up to three months. In 2014 as well as in 2021, the river water

Fig. 3.7 The *Manaus* Opera House was built in 1896 at the height of the rubber boom

flowed in the streets behind the port. Just behind the port terminal is the pretty *Nossa Senhora da Conceição* Cathedral, built in 1877.

Other tourist attractions include:

- The *Bosque de Ciênca* (Science Forest), a large intact jungle in the middle of *Manaus* with huge trees, equipped with mighty butress roots lianas, *epiphytes* (Sects. 6.5 and 5.10) and numerous wild animals such as agoutis, sloths, capuchin monkeys and many bird species, as well as in many enclosures manatees ("sea-cows"), giant river otters, caimans, etc. There is also an interesting museum on this site. The National Fish Institute of Amazonia (INPA) maintains this facility;
- The military zoo, which shows many typical animal species of the Amazon rainforest, although the animals are housed in not very contemporary enclosures;
- The *Mindú Park* (*Parque do Mindú*), a large park-like area with restaurants and an orchid house where numerous animal species are exhibited;
- The *Hotel Tropical* is located directly at the banks of the *Rio Negro* (currently closed). The sightseeing flights over *Manaus* and to the confluence of *Rio Solimões* and *Rio Negro* start from this place. Attached to the hotel is a

small zoo with monkeys, birds, peccaries and a jaguar, and in the immediate vicinity is an intact rainforest, which is home to many interesting birds and insects as well as a group of the very rare pied tamarins (*Sanguinus bicolor*), which are endemic to the *Manaus* area (Fig. 9.10);

- And moreover *Ponta Negra:* this is the promenade of the residents at the weekends, an area with stage, large sandy beach (at low tide), many bars and stalls with typical food of the region; very close to the former *Hotel Tropical.*

3.4 Parintins

The journey continues downstream. The next city worth mentioning is *Parintins*, 470 river kilometres from *Manaus*. It is located on a large island in the middle of the river (*Rio Tupinambarana*) and has 110,000 inhabitants. Although this city can only be reached via the Amazon, there are many motorcycles (mostly as taxis) and cars. The town has attracted attention through many television reports in recent years, as once a year on the last weekend of June it hosts a rousing folkloric festival reminiscent of the carnival in *Rio*. Then thousands of visitors come with their boats, which of course do not all have room at the small pier. So there are many, many boats next to each other, so that late arriving visitors have to jump hundreds of meters from boat to boat until they finally reach solid ground. In a big stadium with space for 16,000 visitors, the *Boi-Bumbá-Show* takes place (*boi* = ox, *bumba* = to hit; thus "hit the ox"). This is a competition between 3500 participants for the most successful presentation of

- figures, for example, animals of the river or forest, such as the *tucunaré* (cichlid), jaguar or caiman,
- fantasy animals, like dragons or giant snakes,
- costumes of the dancers and,
- rhythmic pieces of music.

This competition is always between the same two parties: the supporters of the Blue Ox (*Boi Caprichoso*) fight against the supporters of the Red Ox (*Boi Garantido*). The supporters of these two groups wear blue or red shirts accordingly. The roofs and fences of the houses in the city are also painted blue or red, and the signs of the banks are of course two-coloured (red and blue), so as not to upset the customers. There are even taxis painted blue or red and even Coca Cola joins in: the advertising signs also appear in blue and besides the well-known red cans there are—unique in the world!—also blue ones.

At the next small town, which is passed downstream, the Amazon narrows to "only" 1890 m width. *Obidos* is actually the narrowest point of the entire Amazon, because on both sides of the river there is not alluvial land as usual, but over 90 m high, hard granite rock, which prevents a greater expansion of the river even during the rainy season, so that it comes to this "bottleneck". The Brazilians also call this place *Garganta do Amazonas* ("throat of the Amazon"). Due to this constriction, the river has a strong current and its greatest depth of about 120 m, whereas otherwise it is only between 15 and 30 m. At this strategically important point, the Portuguese conquerors built a settlement in 1697, which was expanded into a fort (*Fort Gurjão*). The town itself has a pretty old quarter with narrow streets and ancient villas.

3.5 Santarém

The next major city downriver is *Santarém*. This large city with 270,000 inhabitants is located in the state of *Pará*, named after the large branch of the Amazon River that leads to the mouth of the Amazon River system into the Atlantic Ocean. With an area of 1.25 million km^2, this state is also huge and more than 3 times the size of Germany or of the state Arizona in the US—with a population of eight million. *Pa'rá* means "river-sea" in the Tupi-Guaroni language. In fact, in this area the lower reaches of the Amazon is becoming more and more like a huge lake or a sea. *Santarém* is situated 51 m above sea level, on the right side of the Amazon—more precisely, where the clear-water *Rio Tapajós* flows into the white-water Amazon River. From this point it is still 970 river kilometres (750 km linear distance) to the mouth. As it was just described at this point there is also a confluence of two different water masses: The clear water of the mighty *Tapajós*, which appears blue from above when the sun shines, flows into the loamy-yellowish water of the Amazon. Similar to *Manaus*, both water masses flow side by side for up to 6 km during the rainy season due to their different flow velocity and temperature before they mix (see Table 2.2). *Santarém* is connected via the road network of Brazil with *Rio de Janeiro*. There is also a well-developed road to the recreation area *Alter do Chão*, 38 km away, where city dwellers like to spend their weekends on beautiful white sandy beaches—in the dry season. This is why this area is nicknamed the "Caribbean of Amazonia" (Fig. 3.8). The inhabitants of *Santarém* live mainly from the wood of the forest, from cattle breeding and from fishing.

Fig. 3.8 In the dry season there are snow-white sandy beaches in *Alter do Chão*

3.6 Belém

Belém—named after the biblical city of Bethlehem—is the last large city at the beginning of the Amazon estuary—still about 120 river kilometres from the Atlantic Ocean. The city is situated only 10 m above sea level on the Bay of *Guajará* (*Bahia do Guajará*), which is part of the *Rio Pará*, a large tributary of the Amazon that has joined the *Rio Tocantins*.

Belém was founded in 1616 by Portuguese sailors as *Forte de Castello* and is now a metropolis of two million people. It owes its economic boom to the rubber boom 120 years ago. Even today, signs of the former wealth can still be seen in the city centre:

- On many house facades there are still old Portuguese tiles;
- The port buildings from 1901 with the *Ver-o-peso market* (in English: "Watch the weight!"). The market is of an overwhelming variety and literally overflowed with vegetables, fruits and fish, medicinal herbs and spices (Fig. 3.9);

- The neighbouring harbour (*Estação das Docas*) has been completely renovated and is now at the weekends an interesting promenade for the local population with shops and restaurants;
- The *Basilica de Nossa Senhora de Nazaré*, built in 1909, is based on the Roman model with columns of rose granite and walls of Carrara marble in the neoclassical style;

Fig. 3.9 The *Ver-o-peso market* in *Belém*

- In addition, the *Catedral da Sé*, built in 1771 in a neoclassical and baroque style, is located close to the harbour right beside the fortress;

Further is worth seeing:

- The *Emilio Goeldi* zoological-botanical garden with a small jungle full of magnificent trees, a pond with the Queen Victoria's water lily (*Victoria amazonica*) and a series of enclosures with rainforest animals. These cages, however, no longer correspond to today's zoo-didactic ideas;
- The *Mangal das Garças* ecological park, opened in 2005, is located near the fortress. This park is home to many native flowering plants and free-roaming birds, as well as an orchid house, a butterfly house and a large bird aviary.

Many streets in the city centre are lined with tall, shady mango trees that are over 100 years old. From the end of April, the mangoes are ripe and fall from the trees in large numbers.

By the way, the city owes its nickname "Belémhatten" (after *Manhattan* in *New York*) to the numerous office skyscrapers.

In order to reach the Amazon upstream from *Belém*, the ships usually choose the passage through the *Brevis channels* (e.g. via the *Rio Parauaú*), past the small town of *Brevis*, and then further on via a series of other natural channels (*Igarapés*) until finally, after a journey of less than a day, the main river is reached.

4

The People of the Amazon River

The Amazon lowlands are predominantly inhabited by *caboclos*, the descendants of European immigrants and indigenous peoples. Both groups are discussed in detail, including their daily lives, how they obtain food, their family structure and their village civil society. It is also discussed how the indigenous peoples have been able to survive in the rainforest until today and how they are currently coping with modern Western civilization. In this context, the traditional community houses (*Malocas*) are presented and those traditional customs are discussed that have survived until today. The following indigenous groups are described: *Yagua*, *Ticuna*, *Huitoto*, *Bora*, *Desano* and *Tupi*. The *caboclos*, like the indigenous people, live mostly in small settlements and villages on the banks of rivers. However, there are also numerous families who live far away from any infrastructure—completely on their own in simple wooden huts built on stilts. In this remoteness they care daily for their own food supply: that is catching fish and cultivating a small plantation.

4.1 The Original Population

According to recent archaeo-genetic studies, the settlement of the Americas began about 23,000 years ago, when the Bering Strait formed a land bridge between the Asian and American continents during the peak of the more than 70,000-year long *Wisconsin Ice Age* and the associated 70 m lower water level. At that time, a first wave of immigration of Eurasian human groups took place across Siberia. The descendants of this population then gradually moved on along the coasts, reaching the southern tip of Tierra del Fuego by

L. Staeck, *Fascination Amazon River*, https://doi.org/10.1007/978-3-662-64452-2_4

14,600 years ago. As genetic studies reveal, members of a second founding population that arrived in the Americas during another cold period are strikingly distinctly related to the indigenous people studied in the Amazon region (Figs. 4.1 and 4.2). These migrants had similar genetic makeup to the people living today in New Guinea, Australia, and the Andaman Islands in the Bay of Bengal (Fig. 4.3). The migrants of the first wave of immigration are more closely related to the indigenous peoples of North and Central America.

At the height of colonization—before the European discovery of the Americas at the end of the fifteenth century—some five to seven million members of indigenous peoples settled in the Amazon basin. How was this possible with this poor rainforest soil known as "climax soil", which is mostly a clay-rich *oxisol* or *latosol soil*? This is a whitish or reddish, depleted, predominantly infertile soil whose humus layer is only a few centimetres. The answer is: In their settlements the indigenous people had created usually 2–3 ha large waste dumps, in which all organic waste (food leftovers, bones, fish bones, plant remains) and own faeces were collected. These waste dumps were mixed

Fig. 4.1 A member of the *Ticuna Indians* in traditional dress

Fig. 4.2 Members of the same ethnic group—dressed in modern clothes, they belong to the clan of unwinged animals, namely the jaguar clan

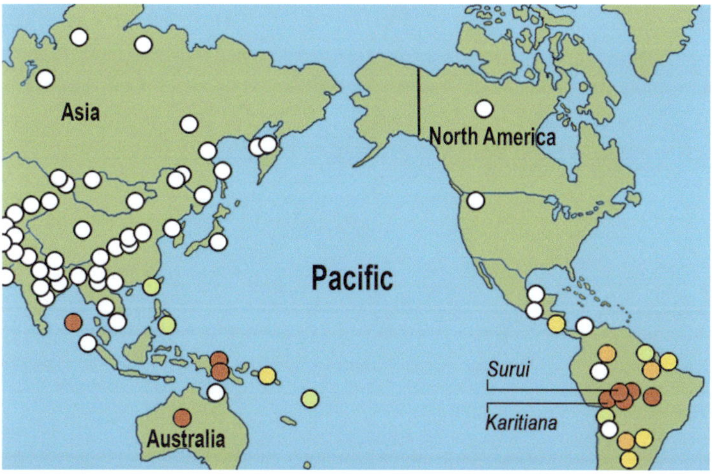

Fig. 4.3 The indigenous population of the Amazon shows great genetic similarities with people living today in New Guinea, Australia and the Andaman Islands—the darker the colour circles, the closer the relationship

with charcoal ash and shell limestone acquired in barter trade. Over time, this soil became a deep black, fertile soil called *terra preta*, which was used as a basis for their gardens and plantations. It is weakly acidic (pH 6.8) and, due to the bone admixtures, contains extraordinarily high levels of phosphate, up to 184 ppm (*parts per million*)—normally fertile soil has 10 ppm—as well as calcium, potassium, magnesium and humus matter. The components of this soil were very stable in terms of decomposition and leaching due to the added charcoal, the shell limestone and the lactic acid bacteria from the dung. Currently, more and more such ancient *terra preta* soils, mostly 20–30 cm thick, are being found in the Amazon basin (Fig. 4.4); in such heaps, which are more than 1000 years old, old ceramic fragments of the cultures of the time have also been found.

When the first Europeans penetrated this area, they brought back reports and drawings of man-eating savages and strange-looking or grotesquely masked people. For example, an image of the *Ewaipanoma* Indians, drawn according to rumors, reached Europe in 1599, showing headless creatures with facial markings on their chests (Fig. 4.5). Other depictions show blood-thirsty Amazons capturing, roasting and consuming white invaders or Indians disguised with animal and demon masks (Fig. 4.6).

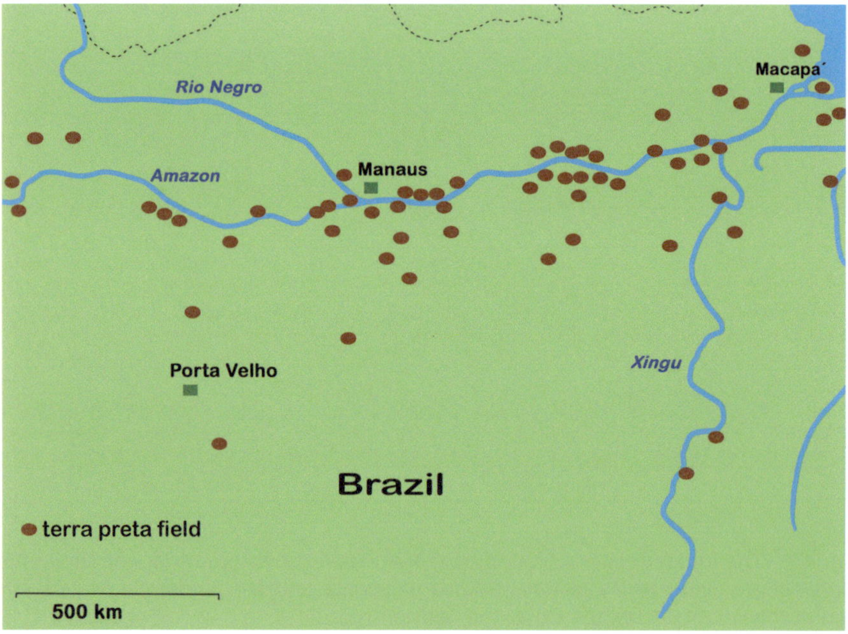

Fig. 4.4 *Terra Preta* discoveries in the middle and lower Amazon

Fig. 4.5 At the end of the sixteenth century, pictures arrived in Europe that were supposed to show members of the *Ewaipanoma* Indians from Amazonia: headless beings with facial drawings on their chests

Due to persecution and introduced diseases, the Indian population has drastically decreased since their immigration. In addition, there has been further decimation due to intermixing of parts of the indigenous population with European immigrants. Exact figures on the current size of indigenous people in Amazonia do not exist. But estimates suggest that there are only about 70,000 indigenous people left in the Brazilian part of Amazonia, living in about 50 different peoples, each with a few hundred members. Of these—according to estimations by the Brazilian Indian Affairs Agency FUNAI—there are currently still up to 20 indigenous groups living in isolation in the rainforest, whose language and habits are completely unknown.

The Indian tribes of today mostly consist of smaller village units that have been pushed into the remotest areas of Amazonia, mostly into *terra firme* forests. Whether or not they have abandoned their original traditions and ways

Fig. 4.6 The indigenous people believed in animate nature. At ritualized festivals, Indians dress up with demon and animal masks

of life is mainly related to how intensively they come into contact with "modern Western civilization" in their habitat. If groups of indigenous people live at least 7 days' walk from a navigable river, it can be assumed that these groups still live largely according to their original customs because of their isolation. Due to the arbitrarily drawn borders in the course of colonization, there were also major resettlement projects of indigenous groups in the Amazon basin, more on this later in this book.

On the direct banks of the Amazon or its immediate tributaries, different indigenous peoples have settled in the past, such as the *Yagua*, the *Ticuna*, the *Huitotos* and *Boras*, and the *Desano*. The members of the indigenous population of Amazonia are still not considered Brazilians. They do not have automatically a Brazilian passport, do not pay taxes and live without the state laws.

The *Yagua* people currently consist of about 6000 members and settle in today's Colombia and northeastern Peru, for example along the Amazon and the *Rio Napo*, spread over about 30 villages. They speak their own language. Their name comes from the *Ketschua* word *yawar*, which means "blood" or "colour of blood". Traditionally, the members of this ethnic group paint themselves with the blood-red seed colour of the lipstick tree *(Bixi orellana)*, which is endemic to the Amazon basin (Fig. 7.8). The impressions and photographs reproduced in this book are based on frequent visits to the *Yagua* village of *Libertad*, Colombia, with a population of about 300, located on the northern bank of the Amazon River, about 110 km west of *Leticia*. The women and men traditionally wear today red-dyed bast skirts and tops made from the leaves of the *Buruti* palm *(Mauritia flexuosa)*, the women also wear red linen skirts, that is, for their festivals as well as when groups of people from another indigenous groups announces themselves (Fig. 4.7). In everyday life, however, they wear Western clothing. They are Christian missionized, but they also retain their animistic beliefs, according to which, for example, natural phenomena such as rainbows, lightning, thunder, storms, but also the sun, the moon and giant trees of the forest are ensouled. Ancestor worship is also part of animism. There is a church and a primary school in the village. Besides many dogs, the *Yagua* keep different, mostly young animals of the rainforest, which are proudly presented to the visitors. Thus, visitors may see animals that are usually hidden in the rainforest and can only be observed during longer stays in this region, such as sloths, potos (kinkajou) *(Potos flavus)*, capybaras, agoutis, pygmy marmosets, squirrel monkeys, woolly monkeys, small red-tailed boas *(Boa constrictor)*, water turtles, small spectacled caimans (yacare caimans), small iguanas, different parrotlets, and giant rhinoceros beetles *(Megasoma acteon)* (Fig. 4.8).

The *Bora*- and *Huitoto Indians* (also spelled *Witoto* or *Uitoto*) are closely related ethnic groups, but speak different languages. They now live in neighboring villages in the border area between Colombia and Peru, for historical reasons. Originally, they lived along the *Rio Putumayo* in Peru. As a result of border disputes between Colombia and Peru, this region became Colombian

Fig. 4.7 *Yagua* women in their typical traditional clothing

Fig. 4.8 Boys of the *Yagua* tribe present their "pets": a parrotlet, a small squirrel monkey and a small spectacled caiman

in 1922. As a result, most of the members of these two tribes left their homeland to settle on the *Rio Ampiyacu,* tributary of the Amazon River, near the village of *Pevas* in Peru. Both tribes suffered tremendously from enslavement during the rubber boom at the beginning of the nineteenth century, so that their populations declined from the original 50,000 to now about 500 *Boras* and 9000 *Huitotos*.

On traditional occasions, they wear skirts and trousers made from the bark bast layer of a fig *(Ficus spec.)* (Fig. 4.9). The black body paint and the applied patterns on the clothing come from the sap of the *genipapo* tree (Fig. 4.22a, b). These patterns have not changed in the last 100 years (Fig. 4.10a, b).

The *Ticuna Indians* with an actual number around 44,000 inhabit the upper *Rio Solimões* in Brazil, Peru and Colombia, with Brazil having the largest number (around 32,000). Like other ethnic groups in the Amazon basin, the Ticunas also suffered from enslavement during the rubber boom. These indigenous people also speak their own language, indicating that they lived in isolation from other ethnic groups for a long time. My impressions and photos of these people go back to frequent visits over the past 25 years, especially to the village of *Vendeval/Brazil* on the banks of the *Solimões* (Amazon). During this time the number of inhabitants has more than doubled, if not tripled, the number of children, especially small children, is enormous,

Fig. 4.9 The clothing of the *Huitotos* is traditionally made from beaten bark bast of a fig species. The upper parts of the body are dyed black with the sap of the genipapo tree

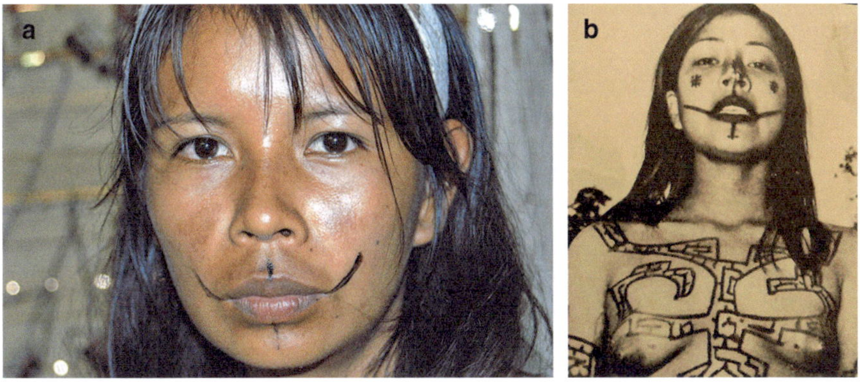

Fig. 4.10 (**a**) Huitoto woman today, (**b**) Huitoto woman around 1870

apparently the *Ticunas* do not practice contraception. The brother of the village chief, for example, has 30 children. Since the inhabitants have had regular contact with white people passing by on (cruise-)ships for a long time, they are dressed in western style and also hardly maintain their traditional customs—with one major exception:

When the girls menstruate for the first time, the great traditional festival of *Pelazon* is held, dedicated to the change from childhood to fertile woman.

Since the number of children is so extremely large, this takes place with corresponding frequency, and several girls are always grouped together for one festival. Two days before the activities begin, the alcoholic drink *masato* is started to be made. For this, cassava flour is chewed and spat into a vessel with the accumulated saliva. With the addition of water and sugar, the liquid ferments within 2 days into a dirty-yellowish, sweetish drink with an alcohol content of about 5–12%. On the feast day of ceremony the concerned girls are stripped naked, bathed in the Amazon, then smeared with the black sap of the *genipapo* tree, and after that their clan symbol is applied. Afterwards they are decorated with pollen, chains from bird feather and a crown also made of bird feathers (Figs. 4.11 and 4.12). Accompanied by music (mainly drums), they are led into the meeting house decorated with leaves and flowers, where the older female relatives (mother and aunts) pull out all their head hair in clumps, with the participation of the villagers (Fig. 4.13). Finally a white headscarf is tied around them as a signal that they are protected from sexual advances by the male village youth for the next 9 months until the hair grows

Fig. 4.11 After the bath the *Ticuna* girl is decorated with *genipapo* juice and bird feathers

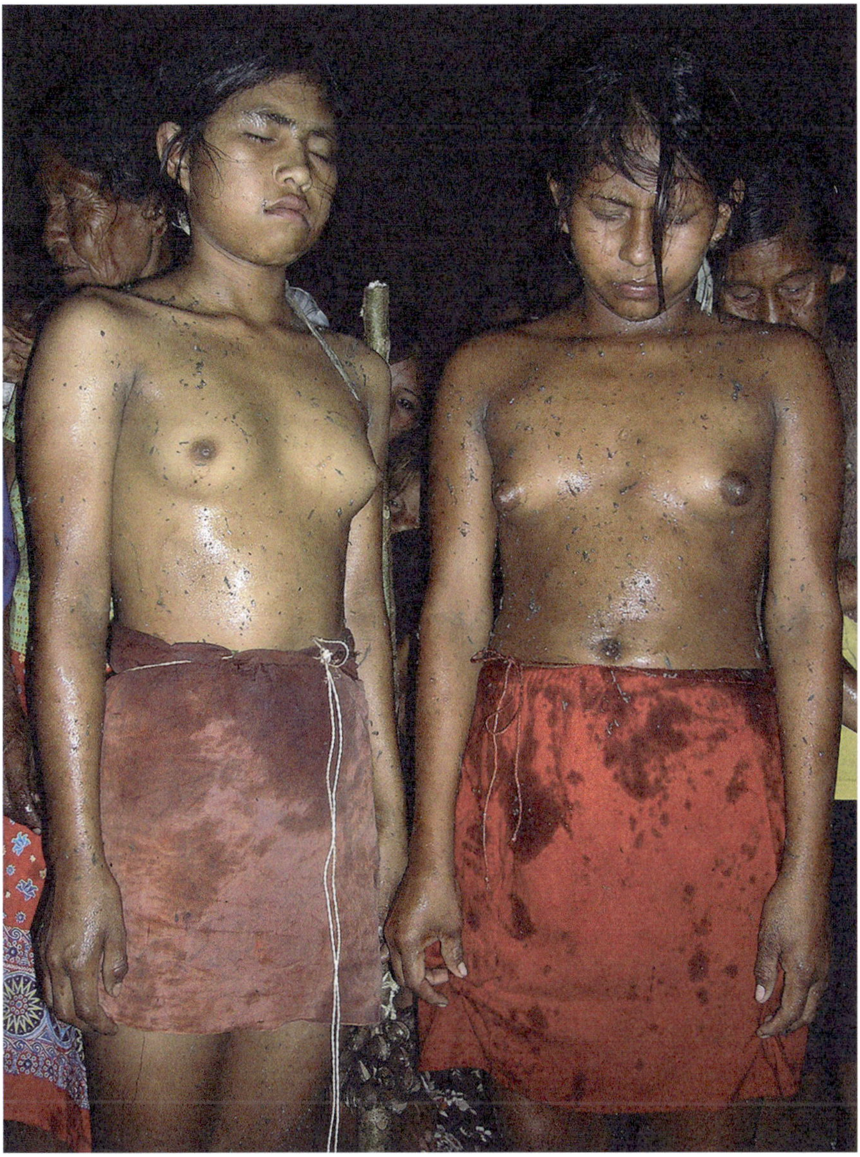

Fig. 4.12 The girls are led into the meeting house ...

back. The festival usually ends with a drinking bout, especially for the male villagers.

In the past a ritual was also performed on boys. When they reached puberty, they also had to undergo an extremely painful procedure: their upper incisors

Fig. 4.13 ... where female relatives pull out their head hair in clumps

of the upper jaw were abraded with simple tools until they resembled piranha teeth in shape (Fig. 4.14).

There is another special feature of the *Ticunas* that is still practiced today: the exogamy law. This allows marriages and actually also sexual contacts only outside of the own extended family, whose members are related to each other by blood. Several extended families form one clan (consisting of different families). This marriage order is supposed to prevent the spread of hereditary diseases. All inhabitants of a village belong to extended family and a certain clan. In the village of *Vendeval*, which I visited, the members of one exogamous clan bear names of winged animals (for example toucan, macaw or eagle) and the members of the other exogamous clan bear names of unwinged animals (for example jaguar, caiman or anaconda). Children become members of the paternal clan. When holidays occur, such as the *Pelazon* festival described above, sexually active villagers must apply the symbol of their clan to their face, i.e. the jaguar clan the tactile hairs of the jaguar (Fig. 4.2), to ensure that the partner belongs to a different clan before engaging in sexual contact.

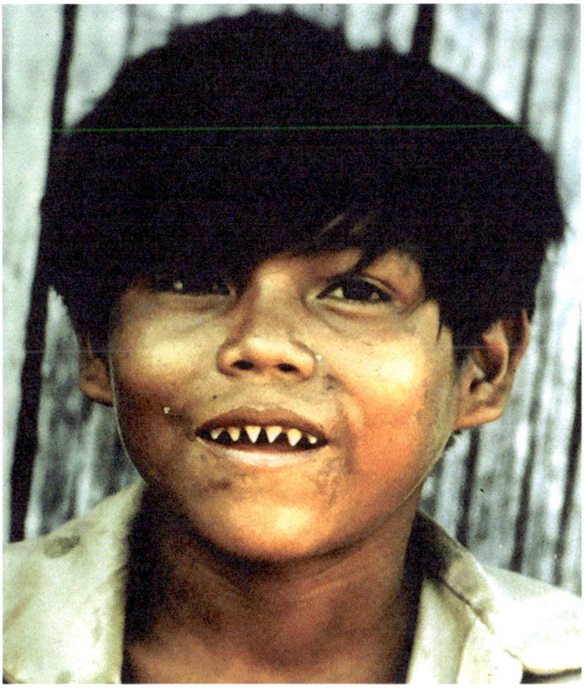

Fig. 4.14 *Ticuna* juvenile with filed teeth resembling piranha dentition

The *Desano Indians* inhabit the northeast of Brazil and parts of neighbouring Colombia on both sides of the *Rio Negro.* The members of this tribe also call themselves *gente do universo* (= "people of the universe"). Due to early contacts with European immigrants and the diseases they brought with them—especially measles—the number of members of this ethnic group was drastically reduced and today amounts to only about 1500 people, spread over 60 communities. This ethnic group also practices exogamy. My impressions and photos of this people go back to visits during the last 15 years in their village *Praia do Tupé,* about 20 km west of *Manaus* upstream on the left bank of the *Rio Negro.* Their traditional dances, body painting and body decoration are impressive (Figs. 4.15, 4.16 and 4.17).

Finally, the important ethnic group *Tupi* must be mentioned as they historically characterize the entire region. This indigenous group inhabited large parts of the coastal region of present-day Brazil and the lower to middle Amazon with a large population of over one million members and numerous different tribes. They all spoke the *Tupi* language, which was also used as a unifying language by other tribes as well as by the immigrant Europeans. Therefore, many terms of daily life of plants and animals known to them still

Fig. 4.15 Dance performance by *Desano Indian* men in traditional dress, with necklace of caiman canine teeth

have names from the *Tupi* language. From the beginning of the eighteenth century, this ethnic group was also largely wiped out by introduced diseases such as influenza, chicken pox, measles, polio or venereal diseases, as well as by enslavement and persecution, so that today only scattered subgroups live in Amazonia.

There are no roads in the Amazon basin. Thus, all traffic is carried out via (water)-roads. Due to the continuously contact of the indigenous population with our civilization, these ethnic groups have adopted many of our behaviors over time, such as religion, clothing, house construction and schooling. Increasingly, there are cell phones and even Western-style television and music. The latter was possible by the project *Luz para todos!*" (="Light for all!") launched by former Brazilian President *Lula*. With its implementation, more and more villages—also inhabited by indigenous groups—have been connected to the basic electrical supply since around 2014/2015—initially free of charge, meanwhile for a certain monthly flat rate. For this, wide swathes have to be cut through the jungle. As a result, the residents complain about increasing numbers of mosquitoes in their huts. In addition, many trees that were originally in the shade are now dying at the border of these lanes, so that

Fig. 4.16 Young *Desano* woman with typical body painting

the corridor is becoming wider and wider. How the necessary free space for the electric lines is maintained in the long term, remains a major logistical task.

Despite all the modern innovations, these indigenous groups have also retained some of their traditional ways of life. For example, although dwellings are increasingly being built with stone walls and roofs made of corrugated iron, the traditional way of building without wires and nails can still be seen: walls made of reed leaves or boards from the trunk of the *Buriti* palm *(Mauritia flexuosa)*, posts made from the trunk of the same palm and the roof covering also made from the large fronds of this palm (Fig. 4.18a). In most cases, the dwellings are also built on stilts, so that the household goods can still be

Fig. 4.17 *Desano Indian* with feather crown made of macaw feathers

stored under the floor and domestic animals can also be kept. In addition, the frequent floods cannot reach the living quarters so easily.

The meeting houses of the *Huitoto* and *Yagua* Indians, for example, are still built today exclusively from plant material from the jungle. These community houses, called *malocas*, are the largest structures in the Amazon, with a ridge height of 20 m and an equally large length (Fig. 4.18b). Because of the way they are constructed, they have excellent ventilation: the air inside is not stuffy, but cool due to a constant light breeze. There is one entrance for men and one for women. In the past, the families of the village lived in this structure in separate units, but today they are used for festivals and ceremonial events.

Fig. 4.18 (**a**) Hut of the indigenous people in traditional construction, (**b**) *Maloca* (community house) of the *Yagua* and *Bora Indians*

The indigenous people keep as farm animals mostly only chickens, pigs are rare in Amazonia, because their nutrition is very similar to ours and thus very costly. They keep as pets very often dogs and furthermore in principle all animals that (still) live in the surrounding forest, for example monkeys, turtles, small boas, agutis, sloths, small caimans (in the cooking pot as a toy for the children) as well as among the parrots mostly the parrotlets.

They grow cassava, sweet potatoes, pineapples, papayas, cacao and beans, as well as banana forbs, yams and taro tubers (introduced from South Asia in former times) in their gardens and plantations. Furthermore they plant different fruit trees for example orange trees and Malay rose apple tree *(Szygium malaccense)*, both introduced, guava tree *(Psidium guajava)*, cashew trees (Brazilian *Cajú, Anacardium occidentale*), sugar apple trees (*Annona squamosa*) and Açai palms (*Euterpe oleracea*) as well as calabash trees *(Crescentia cujete)*, from whose hollowed-out, still unripe, often child's head-sized fruits bowls, ladles, but also masks are made.

Normally, the indigenous people still practice *shifting cultivation*. This means that they clear a limited area of primary forest, perhaps one or 2 ha in size, and then remove the ground vegetation. The trees are cut, but the trunks are left lying around and roots are also left in the ground in order to prevent soil erosion caused by the permanent rainfall and strong solar radiation (Fig. 4.19).

In contrast, caboclos almost always completely clear a large area of forest, so that after a short time the thin fertile soil layer is destroyed by erosion (Fig. 4.20a–c). In addition, *caboclos* often engage in useless slash-and-burn agriculture, which very often gets out of control, causing wildfires. In slash-and-burn cultivation, 90% of the biomass and 40% of the mineral nutrients are lost. In *shifting cultivation,* when the soil fertility is exhausted after a few

Fig. 4.19 Principle of *shifting cultivation:* cleared piece of rainforest, tree trunks and roots left lying around

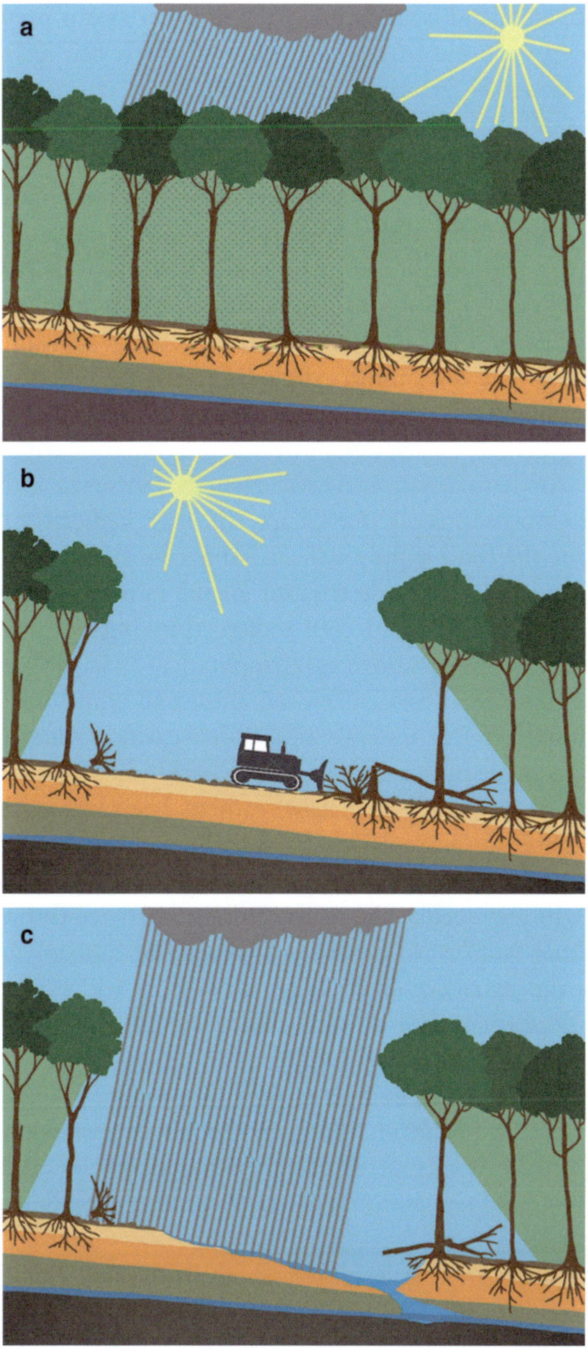

Fig. 4.20 (**a**) Intact forest with a closed canopy cover, (**b**) Trees are completely removed by caboclos and white settlers, including trunks and roots, (**c**) Massive soil erosion takes place after only a short time

years, the cultivated land is abandoned and another area is prepared for agriculture nearby. In the meantime, the primeval forest can reclaim the abandoned land in about 5–10 years.

From the neighboring rainforest, the residents get a wealth of useful plants. While we city dwellers only see "green" in the jungle, the indigenous people distinguish. For them, the forest has a great value: it is not only a grocery store (for meat, eggs, vegetables, side dishes, fruit, etc.) and a construction market (wood, leaves for building houses), but also a pharmacy (herbal medicines). Finally, the river itself is vital for its inhabitants: it is not only a supermarket (fish, crustaceans, turtles), but also a bathtub, toilet and washing machine. If you add the enormous amount of edible fish from the Amazon rivers to the products from the forest and to the vegetable and fruit varieties cultivated in plantations and gardens, it becomes obvious that the food supply of the people in the Amazon basin is currently still ensured. However, as the population figures continue to rise steeply (see Chap. 10), it is predictable that in one or two generations the food supply will no longer be sufficient to feed all the inhabitants—especially as the destruction of the rainforest continues unlimited. A change in political and civil society thinking is therefore urgently needed in order to avoid a human catastrophe.

Under the influence of missionization, almost all ethnic groups now wear Western clothing; only at festivals traditional garments are worn, such as shorts and skirts made of beaten bast (paper bark) of certain species of *Ficus*, feather crowns, mostly made of macaw feathers, necklaces made of the seeds of different plants (for example, the açai palm, its seeds are dyed), the red-black seeds of the rosary pea *(Abrus precatorius)*, imported from India, the arrow grass (light grey seeds), the seeds of the rubber tree (brown-black marbled seeds, *Hevea brasiliensis*) or of teeth (mostly of caimans, boars or dogs, more rarely of the jaguar) or of the huge, 4 cm long, white bleached scales of the pirarucu *(Arapaima gigas)*. Such chains are also sold as souvenirs for tourists.

Amazonian animals (such as the anaconda, freshwater dolphin or pirarucu) are carved from the beautiful, blood-red wood of the muirapiranga tree (*Brosimum paraense, Moraceae*), also known as bloodwood, red rosewood or false brazil tree (*falso pau brasil*) (Fig. 4.21) or from the light wood of the balsa tree.

Mats, skirts, ropes and baskets are made from the bast of the Buruti palm and from various lianas.

The two important colours for applying the traditional colour patterns are red and black. The already mentioned bright red dye *urucú* (*rucu* = "red" in the *Tupi* language) is extracted from the seeds of the lipstick tree *(Bixi*

Fig. 4.21 Wood of the muirapiranga tree, also called bloodwood

orrellana) (See also Fig. 7.9). The indigenous people use these seeds to dye clothes, but also food and skin and hair. This, by the way, is how the name "redskin" came about.

The black dye is obtained from the crushed pulp of the genipapo tree (*Genipa americana*; plant family *Rubiaceae*), which grows in the terra firme. In this process, humid pulp of the still raw fruit is first pulverized, after this the loose plant tissue is pressed through a sieve, and finally the clear plant juice is collected and applied to the skin, for example with a bamboo tip (Fig. 4.22a, b). After about 10 min, the sap oxidizes and turns black (Figs. 4.9 and 4.10). The patterns look like a tattoo on the skin, but only remain visible for about 2 weeks before they peel off. However, the applied sap is also used as an insect repellent. The ripe fruits are also popular for making fruit juice, they contain vitamin C. The word *genipa* is also borrowed from the *Tupi* language and means "breast of an old woman".

For the inhabitants of the rainforest, the immense abundance of plants offers a variety of plant secondary metabolites: drugs, medication and poisons to stun or kill their prey. In the course of their evolution, many plants have acquired biochemical properties that protect them—as they cannot run away—from phytophagous (= herbivorous) predators. Thus, they have stored a wide variety of chemical substances (for example alkaloids). Medicine men and women, healers, plant specialists and shamans within the village

Fig. 4.22 (**a**) Fruits of the genipapo tree, (**b**) Prepared and oxidized juice of genipapo fruits

communities have taken advantage of this for many generations and use many plant species as healthy food supplements as well as for the treatment of diseases and injuries. For example, there are medicines for headaches, nausea, diarrhea, worm infestations, bleeding wounds, abscesses, and even contraception and abortion. For instance, the small shrub pião (or *pinhão-roxo, Jatropha gossypiifolia, Euphorbiaceae*) with its pretty, reddish leaves should not be missed in any garden. An extract from the leaves is considered an old Indian home remedy. The sap is mainly anti-inflammatory and haemostatic and is used against various diseases, for example nausea, diarrhoea and abdominal pain. This is why this shrub is also called *bellyache bush* in English (Fig. 4.23). The shrub *camu-camu (*Brazilian *araçá de água, Myrciaria dubia, Myrtaceae* family), which grows up to 5 m high and likes to grow on the sunny shores of blackwater lakes and rivers, also plays an important role in the diet of the

Fig. 4.23 Inflorescence of the piaõ bush

Fig. 4.24 The berries of the *camu camu* bush have 40 times the vitamin C content of oranges

indigenous population. It has small white flowers that ripen into reddish fruits between December and April. In appearance they resemble cherries but are larger (Fig. 4.24). They taste sweet and sour, are very rich in iron, and have an exceptionally high vitamin C content (up to 3% of fresh weight), 40 times that of oranges! In addition, their consumption lowers blood sugar and blood lipids. Among the locals, they are also considered as an aphrodisiac.

But not only the healers, also the other villagers usually cultivate important medicinal plants near their huts in addition to kitchen herbs. Among the most important diseases of the indigenous people are malaria, tetanus, diarrheal diseases, respiratory infections, amoebic and worm infections (for example hookworms, intestinal flukes or tapeworms). In addition, Europeans introduced TB, rubella, chickenpox, venereal diseases and viral influenza.

The indigenous people also use the poison of the bush rope/devil doer *(Strychnos toxifera)* to prepare their arrowheads for hunting animals. It contains an alkaloid in its seeds known as *strychnine*. In smaller doses it stuns humans, in stronger doses it kills them. The plant is a climbing shrub growing on the banks of blackwater rivers with shoots that have been transformed into hooks (Fig. 4.25).

A total of several hundred plant (and also animal) poisons are known under the collective name *curare*. They can heal, but also kill. *Curare*, for example, causes muscle paralysis via the alkaloid toxiferin (Fig. 4.25), which leads to respiratory arrest. It acts via the bloodstream and not via the digestive tract. Alexander von Humboldt had already recognized this when he saw Indians eating wild boar killed with *curare*. He even tried *strychnine* and put a small amount in his mouth, which kills a pig within 6 min. Afterwards he wrote in his diary, "It tastes pleasantly bitter, but one should not have bleeding gums!" Once he had accidentally soaked a sock with *strychnine*. He was about to pull it over his foot, which was scratched with flea bites, when he noticed this and thus narrowly escaped death.

4.2 The caboclos

Besides the indigenous population, there is another group of people living in Amazonia: they are *caboclos,* the descendants of Indians and European immigrants. This word is also borrowed from the *Tupi* language. The term derives from *kaa'boc* and means "coming from white people". The faces of the *caboclos* are not as Asian as those of the pure indigenous population, the eyes are not almond-shaped, the skin colour is brown or light brown, the hair colour is consistently black, and they are small, barely over 1.60 m tall (Fig. 4.26a, b).

Most of them live in the metropolis of *Manaus* and in small settlements or villages on the banks of the Amazon and its many tributaries, for example in *São Paulo de Olivença, Amaturá, Fonte Boa, Uarini, Alvarães, Tefé, Coari, Codajás, Anori* or *Manacapuru*. In addition, many individual *caboclo* families

Fig. 4.25 The nux devil doer *(Strychnos toxifera)* contains the alkaloid toxiferin in its seeds, which is highly toxic

live far from any infrastructure on the river, where they build a simple hut on stilts and, in addition, a jetty to which they can tie their canoe—mostly today with a small outboard motor (Fig. 4.27). The huts are made of simple boards, there is no window glass, but wooden shutters. The people obtain their drinking water from the rain. Mostly they have many children (5–10 are not uncommon!), they do not know contraception. Behind the hut they have cleared a piece of rainforest, but they have burned the tree trunks and roots and do not leave them as erosion protection, as the indigenous people practice. It has already been mentioned: the nutrient loss from this slash and burn is immense—90% of the biomass nitrogen and 40% of the mineral nutrients are lost. After the clearing, they plant some banana forbs, maybe some fruit trees (cashew and a guava tree), and cultivate sweet potatoes and cassava. After living and cultivating for five years a new plot of land, *caboclos* acquire a

Fig. 4.26 (**a**) *Caboclos* are descendants of Indians and European immigrants, (**b**) The faces of adult *caboclos* are brown, the hair colour is consistently black

Fig. 4.27 A *caboclo* hut surrounded by a palm plantation (açai palms)

permanent usufruct over it, which they may then sell. It is planned that after 10 years they will be able to register their land as their property in the land registry.

Manioc (cassava) originates from South America. The name cassava goes back to the *Tupi* language *(mani-oca)* and means "house of the Mani". Other common names are *tapioca* (which is pure cassava starch in small white balls) or *yuka* (in Spanish-speaking countries). A weakly woody shrub with a height of over 1.50 m grows from individual shoots planted in the ground. As this plant species belongs to the spurge family (*Euphorbiaceae*), all parts of the plant contain a milky sap which is poisonous. The palmate leaves resemble chestnut leaves (Fig. 4.28). The flowers are inconspicuous. The plants are easy to care for, growing well in tropical climax soils, but need plenty of sun and high temperatures. The spindle-shaped cassava tubers, formed from the thickened lateral roots and weighing up to several kilograms, are harvested after 6–36 months (depending on the variety). Freshly harvested tubers spoil after only a few days, but in the soil they can survive for up to 4 years. They provide people with a starchy flour. Eaten raw, cassava with yellow flesh is poisonous because it contains the glycoside *linamarin*, which releases a derivative of hydrocyanic acid (prussic acid derivative) when peeled.

After peeling, the tuber—containing yellow flesh—must be watered for 2 days. During this process, the hydrocyanic acid is released into the water by fermentation (Fig. 4.29a). The tubers, which are then soft, are pressed through a basket weave (Brazilian name is tipiti) (Fig. 4.29b).

The yellowish liquid that emerges, mixed with prussic acid residues, is often collected, flavored and sold at the market in plastic bottles as *tucupi*. By adding it to soups and sauces, the prussic acid is eliminated during cooking. The moist cassava mass is then roasted on a large pan (Fig. 4.30).

Fig. 4.28 In cassava, the spindle-shaped tubers develop from the lateral roots. In the background, the hand-shaped leaves can be seen on the shoots

Fig. 4.29 (a) After peeling, the cassava tuber must be soaked for 2 days, during which the prussic acid contained in the tubers is released into the water, (b) After soaking, the soft cassava tubers are pressed through a basket weaving *(tipiti)*

Fig. 4.30 The moist cassava mass is roasted on the pan to make flour

Depending on the temperature and duration of this process, fine yellow manioc flour (Brazilian *farinha*) or crispy flour particles are produced, which are served as a side dish with meat or fish.

On the market, the manioc leaves, which are fermented for up to seven days—by which time the hydrocyanic acid has disappeared—are often sold as vegetables *(macaxeira)*.

Meanwhile, cassava has also been cultivated, which is free from prussic acid. The flesh of this variety is white.

The *caboclos* often establish a small plantation of açaí palms *(Euterpe oleracea)*, which easily sprout from seed and grow quickly. This very slender palm, up to 20 m tall, with a canopy of leaflets similar in appearance to the coconut palm, grows only in the Amazon basin (Fig. 4.31). Its fruits, which are only 12 mm in size, and its palm hearts fetch good money in the market. When the palm is fully grown, the last metre of the palm top can be cut off and debarked. The young marrow of the palm that is now released is used as palm hearts. It tastes like fresh asparagus. The topped palm, however, dies because it loses its vegetation point from which the trunk, leaves and ultimately the flowers develop. However, this is not a disadvantage in the case of the açaí palm, as this type of palm is a botanical peculiarity: it is "shoot-borne". This means that there are numerous reddish adventitious roots at the base of the trunk. These are capable of dividing and growing. Thus, the cut trunk sprouts daughter shoots from these roots, which grow into new palms.

Fig. 4.31 The slender açai palms, which grow to a height of 20 m, only grow in the Amazon basin

Even more significant with this palm, however, are its fruits. After harvesting the many kilograms of fruit (Fig. 4.32a), the juice is squeezed out of the flesh of the fruits, which look like blueberries in size and colour and are only a few millimetres thick (Fig. 4.32b). The remaining seeds are discarded. The juice contains phytochemicals, which include not only a number of important vitamins such as A and C and minerals such as calcium and iron, but also *anthocyanins,* which have also been shown to prevent aging of the body (anti-aging properties). Its protein is even similar to milk protein. For a liter of fruit pulp, the *caboclos* receive up to 8 US-Dollar! These palms are therefore reverently called the "green gold of Amazonia". The açaí juice has an earthy, slightly astringent taste of its own, which at first takes some time getting used to. This juice has always been very popular among the entire Brazilian population, especially the inhabitants of cities prefer to buy it as an "energy drink" with water ice and sugar or mixed with lemonade, or they also like to consume it as ice cream.

Meanwhile the cosmetics industry produces a fragrant body lotion from the açaí juice. In the United States as well as in Germany the first suppliers offer refreshment drinks as a healthy "superfood".

Indians or *caboclos* who have drunk açaí juice all their lives actually show hardly any wrinkles in old age and their hair only turns grey in late old age. Not without reason this fruit is also called *fruta, que chora* (= "fruit that heals").

Caboclos hunt basically all animals and birds of the rainforest for their own food, for example monkeys, peccaries, sloths, turtles. If they have enough food, they also care for these animals temporarily as pets. However, they also keep some chickens, rarely pigs, as these require elaborate feeding. All domestic animals are accustomed to high water and then spend the day on jetties during this time. However, the riparians get their basic supplies from the river

Fig. 4.32 (**a**) The fruit stand of the açai palm with fruits that resemble blueberries in size and colour, (**b**) Açai juice

every day. Fish are more abundant in the Amazon rivers than in any other river in the world. Biologists have now described more than 3000 different species of fish. For comparison: in Central Europe there are about 500.

Some *caboclos* even keep zebu cattle (Indian humpback cattle). In Amazonia one cattle needs an area of about 2 ha a huge area for its nutrition. In the US or in Germany, on the other hand, it is only 0.8 ha. In South America cattle farming is therefore with problems; in addition, the cattle hooves destroy the pastures after a few years: after that, the soils become infertile forever due to the onset of erosion. In South America, where there have never been animals with such hooves, there are no hoove-resistant grasses.

Some floating shops on boats that sail along the banks of the Amazon supply the *caboclos* families, living far away from settlements. In the remote settlements there is even sometimes a pharmacy, which is surprisingly well equipped with drugs against malaria, nausea, pain, worm infestation and fungal infections, as well as acid blockers, aspirin and anti-diabetics. Since school is compulsory in Brazil, there is also a primary school in every *caboclo* village.

The village headman—the chief among the indigenous peoples—is called *Kazike*. He is elected by his villagers. There are also churches. For financial reasons these usually consist only of an open, roofed room with an altar. The practiced faith can be quite different and is based on the missionaries who were the first in the region. Thus, there are Roman Catholics, Baptists, Presbyterians, Evangelists and others in the Amazon lowlands.

5

The Rainforest

The tropical lowland rainforests in the Amazon basin are among the oldest forests on earth, dating back some 50 million years. The ice ages coming from the north have never reached the Amazon region and thus have not been able to interrupt the evolution of these forests, as has happened more frequently in Europe, for example. There have only been prolonged periods of drought that have temporarily affected the lush vegetation. The plants of the tropical rainforest depend mainly on three factors: light, warmth, and moisture. The more favourable these conditions are, the more luxuriant their growth and the more diverse their species spectrum. This is why we find such a great abundance of life in the vast Amazon rainforest.

5.1 Greatest Diversity of Plant Species

The most striking feature of the Amazon rainforest is its *diversity of plant species*. It is a *primary forest* or *virgin forest*, i.e. an original forest that has not been influenced by humans and is in its final or mature stage (climax stage), which is reached after about 200–400 years. An intact primeval forest is a highly complex ecosystem with many changing interdependencies and includes all the stages of development that the forest has gone through in the course of time. Thus in this forest you can find

- pioneer woody plants,
- furthermore, the tree species of medium height and moderate thickness that follow with increasing shade,
- and finally the tall hardwoods of the climax species.

L. Staeck, *Fascination Amazon River*, https://doi.org/10.1007/978-3-662-64452-2_5

During the ongoing *succession* of different plant groups or plant communities the species diversity and vertical complexity of the rainforest constantly increase on the long way to the climax stage. For example, a 100- to 150-year-old forest hosts only half as many birds, mammals, and trees as a mature forest. At the end of this development, the primeval forest is in a dynamic, self-regulating equilibrium of food chains and food webs, in which, for example, gaps torn into the forest by dying trees or by a thunderstorm are quickly filled again by young trees from the group of pioneer woody plants that have been holding out for many years in a "waiting position". The primeval forest in its final stage (the climax stage) is composed of a mosaic of *microsuccessional* patches of different ages. All this can be observed very beautifully if the visitor walks attentively through the rainforest.

How species diversity develops from the first pioneer stage through the various ecological *succession stages* to the mature rainforest after up to 400 years is very well reflected in the following summary, which is the result of many years of research work made available to me by the National Institute of Amazon Research INPA in *Manaus* (Table 5.1).

This makes aware that *diversification*—i.e. variety—is virtually the key word in understanding this forest: of the slightly more than 60,000 tree species identified worldwide (status 2017), 6727 (status 2017) have been identified in the Amazon rainforest. Most of them belong to the *legume* (bean) family (*Leguminosae*). The enormous richness of species can also be expressed in this way: on 1 hectare of rainforest near *Manaus*—that is 100 by 100 m—476 different tree species have been identified; in addition, there are many epiphyte and liana species. In contrast, in Germany with 357,000 km^2 there are only distinct under 100 different tree species, which means that in Amazonia there are more tree species on 1 ha than in the whole of Europe! This enormous abundance of species means that within a primary forest in its climax stage only a few individuals occur—often only one specimen of a single

Table 5.1 Development of species diversity on an alluvial area of 5000 m^2 (0.5 ha) in Amazonia (tree species with a diameter of at least 10 cm at breast height)

	Pioneer stage (after 3–5 years)	Early ecological succession stage (after 30–50 years)	Late ecological succession stage (after 100–150 years)	Mature forest (climax stage) (after >300 years).
Birds	21	49	127	236
Primates	0	2–6	6–8	8–12
Trees	19	33	50	112

species within a wide radius. The second striking feature of the Amazon rainforest is thus the *low population density of different species.*

5.2 A "New" Primary Forest Cannot Be Planted

Regarding these striking features of the primary forest it is obvious that humans cannot plant such a "new" forest if the original one has been cut down or burned. A new planting can only be a "secondary" forest, which is significantly poorer in species and is mainly composed of a few fast-growing and sun-loving pioneer trees as well as species introduced from other regions. A secondary forest has a high number of individuals of fewer plant species, which of course has many negative effects on the original animal biodiversity. New estimates suggest that in Amazonia an area of 30,500 km^2 of primary forest is currently being lost in Amazonia every year, which corresponds to the area of Belgium or, in 12 years, to the area of Germany or of Arizona in the US. Reliable figures on global rainforest deforestation are rare. The *Living Planet Report* of the environmental protection organization WWF published a report in 2018, according to which around 239 million hectares of rainforest have been destroyed worldwide since 1990—an area more than six and a half times the size of Germany or of Arizona in the US. In relation to Amazonia, projections assume that around 90% of the remaining rainforest will have disappeared by 2050. Along with the primary forest, all significant animal species will also disappear, as the food chains will collapse.

And what is almost worse for humans: the rain that regenerates itself from the evaporation of the forest—that is half of the total precipitation—is missing forever. How can this phenomenon be explained? The reason is the lack of gaseous *terpenes*, which are produced in an intact rainforest in large numbers by the jungle trees at the beginning of the daily sunrise. These *terpenes*—chemical exhalations of the trees—form at first natural *aerosols*. These particles—with a diameter of about one nanometre—float in the air, combining with the oxygen in the air, forming various other compounds, the so-called *secondary aerosols*. Above the closed tree canopy of the original Amazon rainforest, each litre of air contains half a million *aerosols*. The rising moisture in the air condenses on these *aerosols* into tiny water droplets, which fuse together over the following hours, forming more and more clouds as the day progresses, until they become so heavy that they rain down on the earth. Without these *aerosols*, rain clouds cannot form.

The second half of the total precipitation transported by the trade winds no longer seeps temporarily into the jungle soil because of the increasing deforestation. That is why the amounts of rain immediately return to the rivers, from where they flow without benefit in the direction of the Atlantic.

The richness of species in the primary forest is mainly due to historical reasons. A basic ecological principle is: the longer a habitat has had similar environmental conditions, the more balanced and stable and thus the more species-rich it is—as in this case. The harsh, ever-changing climate with frequent natural catastrophes, such as in Central Europe, leads to a high rate of species extinction, so that only a relatively few different species remain. These, however, have high numbers of individuals for "survival reasons". The enormous number of species in the Amazon rainforest, on the other hand, leads to constant competition, which can only be countered by a high degree of specialisation. In the course of millions of years, even the smallest niches in the overall structure of the rainforest are occupied, whereby hardly any larger numbers of individuals can be achieved; for example, a tree species with only one specimen on 5–10 km^2 often stands alone. Moreover, the enormous number of *phytophagous* (plant-eating) insects also contributes to the fact that only a few offspring of an individual tree species have a chance of growing old. The isolation of a plant species is therefore a survival strategy, predators simply overlook the plant in the jungle lushness. Nevertheless, this example illustrates the great fragility and endangerment of this ecosystem, because even by cutting down or burning small areas, a large number of different species are irretrievably lost.

5.3 The Ground Layer of the Rainforest

The best way to show the incredible biodiversity of the rainforest is in its vertical structure. Upon entering the forest, the visitor quickly notices that a variety of strategies are applied in the great competition for light and space.

Tip

Preparation of a jungle hike

- Sturdy, ankle-high shoes are important because the ground is uneven, roots stick out and it is usually muddy.
- A long-sleeved shirt and long pants will protect you from bloodsucking insects and the long thorns of many plants.
- Rub mosquito repellent on face, neck, arms and legs.
- Take a machete with you to cut branches and vines that block the way. This will also avoid direct contact with plants where biting ants or snakes can lurk.

With the low light yield on the jungle floor of only about 1–2% of the sunlight, there is naturally neither an extensive herb layer nor a significant shrub layer as in many of our forests. Even when the sun is shining, the incidence of light at eye level is low, about the same as 40 min before sunset as in our northern latitudes. There is also no wind on the bottom floor of the forest. It is usually easy for human visitors to get through the vegetation, and the use of a machete is rarely necessary. Predominant in this lowest floor are a wide variety of seedlings and slender saplings of germinated tree seeds up to 1.5 m tall (Fig. 5.1). These have stopped growing. Rather, they persist in the twilight of the lowest forest floor in a kind of waiting or dormant position that can only end when sunlight reaches the saplings directly.

It is uncertain whether this will ever happen, because a clearing or breach in the rainforest can only occur if, for example, a huge, age-old tree dies, taking several neighbouring trees with it. The tree mortality rate in the Amazon rainforest is about 1.7% per year, which means that about ten trees plus a number of neighbouring trees die per hectare in 1 year. By the way, volatile substances in the form of chemical messengers escape from dying old-age trees. As soon as these substances reach neighbouring trees and the saplings in the waiting position, they trigger stress in these plants, which puts them into a kind of "active waiting position". This enables them to grow quickly into the

Fig. 5.1 Dwarfed remaining tree saplings (referred to by me as "oskars") at the bottom of the primary forest

gap that will soon be created. I have named such dwarf trees in (permanent) waiting position "oskar" in reference to the dwarf *Oskar Mazerath* in the novel "The Tin Drum" by the German novelist Günther Grass. In this lowest forest floor grow besides these Oskars, small shrubby palms, here and there a shrub or dwarf tree and a few flowering plants adapted to diffuse light conditions.

A very special perennial plant is adapted to these diffuse light conditions in the Amazon rainforest: The sore-mouth-bush (*Psychotria poeppigiana* nowadays named *Palicourea tomentosa; Rubiaceae*). Its inflorescence features the red-yellow colour combination which is common in the tropics, as this colour mix attracts butterflies (and birds). Once the fruits have ripened after pollination, they present themselves in bright blue (with red), a colour that is more common in fruits but very rare in tropical flowers. Presumably this is a double system of signals: first, red-yellow signals the nectar source to the pollinator, whereas blue-red signals a food supply to fruit-eating birds (for example, *tanagers*, Sect. 9.6) (Fig. 5.2a, b). This double system has advantages for all involved animals like butterflies, birds, no deception is involved. The visitors of the flower save time and energy, the marked targets are immediately recognised, pollination is guaranteed. The same applies to the fruits: the birds recognise the clearly contrasted search pattern, which guarantees the dispersal of the seeds. Before the inflorescence opens to reveal the small yellow tubular flowers, the bright red sepals look like strikingly made-up lips, which is why they are nicknamed "Labias da prostituta" in Brazil (Fig. 5.3).

Mosses can also be found on tree trunks, dead wood or more rarely directly on the ground, but altogether few species. Visible mushrooms, so called "higher fungi" (*Basidiomycetes*), are rarely seen, mostly they grow on dead substrate, for example, on decaying tree remains or leaves. Among the fungi, however, there are spectacular species, for example those that glow at night,

Fig. 5.2 (**a**) The red-yellow flower colour of the sore-mouth-bush *(Psychotria poeppigiana)* signals a nectar source to butterflies and birds, (**b**) The blue fruits of the same plant attract birds in the dark jungle

Fig. 5.3 The bright red sepals of these plants have the nickname "Labias da prostituta" in Brazil

Fig. 5.4 Mushrooms (*Basidiomycetes*) with colourful fruiting bodies recycle leaves ..

emitting a green-fluorescent cold light and thus attracting insects, or almost aesthetically pleasing ones with colourful fruiting bodies (Figs. 5.4 and 5.5).

Among the few plant species that grow in the shade near the ground, are isolated herbaceous plants, most of which have dark green or almost blue-green leaves. The blue colouring favours the filtering out of yellow light, which is important for the process of photosynthesis. The dark green leaves contain

Fig. 5.5 ... and rotten wood; this picture shows a group of the species *Cookeina speciosa*

many more *chloroplasts* (photosynthetically active cell organelles) than leaves in sunlight, so they can become photosynthetically active in the shade. Then there are velvety-feeling leaves that have peculiar coloured or white-ivory injections, spots, stripes or even white dots with line patterns—so as representatives of the genera *Begonia*, *Calathea* and *Maranta*. Today we know that such colour patterns are not a "luxury formation", but they blur the leaf outlines, so that such leaves are more difficult to identify by leaf-eating insects in the dim light on the forest floor. Such patterns are called *mimicry* (deception by imitation) and are used for camouflage. They send out signals that simulate a certain appearance, namely that the leaf is already old and/or has already been eaten off. Insects, which are only capable of blurred vision, fall for this and do not eat from the leaf, but continue to fly (Fig. 5.6a, b). Since these plants look pretty to us humans, they can now even be bought in well-stocked garden centres.

Other shade plants have developed warty, transparent epidermal cells; these are located in the uppermost cell layer, where they act like magnifying glasses and focus the little incident sunlight, making photosynthesis possible. There are also shade plants in which the underside of the leaves is coloured purple to dark red by embedded *anthocyanins*. These pigments reflect the incident residual light up to the cell layer that carries the chloroplasts. They thus double the amount of light and enable photosynthesis.

Only after the fall of a huge, age-old tree or also after slash-and-burn clearing a dense herb layer is able to develop in the resulting clearing, for example

Fig. 5.6 (**a**) The leaves of a begonia *(Begonia nasoniana)* pretend to be old and eaten away, (**b**) In representatives of the *Calathea* and *Maranta* species (here *Calathea elliptica*), the leaf contours are blurred by coloured sprinkles to deceive leaf-eating insects

sun-loving flowering plants (e.g. heliconias, ginger species), shrubbery and sparse secondary forest of fast-growing pioneer woody plants. However, in the course of *succession*, this colonization phase is terminated after a few years by emerging vigorous tree species, and the clearing is overgrown. A similar picture as in open clearings can also be seen in gaps in the interior of the forest, in the root, trunk and crown area of fallen primeval forest giants as well as along the river banks. Here, however, *succession* does not occur because of the persistence of strong sunlight. Reports about the "impenetrable jungle", for example by Alexander von Humboldt in his diaries, are therefore wrong; the impressions that led to these reports were gained from on board a boat, as Fig. 5.7 shows.

In the lowest floor of the rainforest are almost no sounds. From time to time a bird sings, mostly not very melodious, but with loud, fast following sounds, for example the toucan. Here and there also a group of monkeys calls, which announces that this is their district. Particularly impressive are the noisy monotonous concerts of the cicadas, which swell out of nowhere. They are pure male choirs, because only the males of these insects are gifted with sound. By the hundreds, they stroke the edge of their wings simultaneously with their dorsal leg, producing a constant loud sound that breaks off abruptly after a few minutes. This "song" can sound like the loud noise of a passing train or a shrill unvarying sound of a circular saw and can repeat every 10 or even every 30 min for a while.

Fig. 5.7 The river bank, densely overgrown with pioneer woodland, gives the (false) impression that the primeval forest is impenetrable

5.4 Middle Floor of the Rainforest

The second, most significant floor of the jungle is formed by the 20–40 m high deciduous trees, which end with an almost uniform canopy. This canopy can be seen particularly well when flying over the rainforest by plane. The canopy itself is flooded with light, but allows only very little sunlight arriving the ground. There are no conifers at all. The different shades of green of the foliage are particularly striking, whereas flowering trees are rarely visible from the ground. The reason: in the tropics there are no distinct seasons like in the northern hemisphere, especially the winter is missing. The annual spectacle of nature reawakening in Central Europe, North America does not exist in the tropical forests. There one distinguishes the dry season with less rain from June to October and the rainy season with much more rain from November to May. Thus, trees as well as herbaceous plants do not flower at the same time, but each species has its individual flowering cycle. During the flowering phase, many forest trees shed their leaves to make it easier for potential pollinators to spot the flowering splendour. Also the change of leaves does not happen at the same time in all trees, but instead each species has its own rhythm; this can be with one species which change every 10 months its leaves, with another species only every 3 years. Tropical trees therefore do not have sharply defined annual rings. So if you drive along the river by the rainforest, you see all our four "seasons" side by side: one tree is leafless ("winter"), another is just breaking bud ("spring"), another is just flowering ("full spring"), one is hanging full of fruit ("late summer"), another has all its leaves just yellow or brown (-red), ("autumn").

Due to the lack of wind inside the primary forest, almost all plants need animals for their pollination/fertilization and for their seed dispersal. Thus, mainly insects (beetles, butterflies), but also birds and bats are attracted, which consume pollen and nectar or seeds and fruits. Thus, *cauliflory* is quite common in the rainforest (Fig. 5.8). Flowers and thus later fruits arise directly from the trunk. This makes it easier for bats to find them in the midst of the leafy forest.

Only the tiny spores of the flowerless mosses, ferns and fungi are spread by the slightest air currents and then germinate in cracks in the tree bark and on branch forks.

A special phenomenon in the primeval forest is the so-called "sudden unfolding of foliage". Under optimal weather conditions—for example when the rainy season begins—whole bunches of fresh, wilted-looking leaves

Fig. 5.8 *Cauliflory* is quite widespread in the rainforest: this picture shows wild cacao *(Herrania mariae)*

Fig. 5.9 Fresh young leaves hanging limply to fool leaf-eating insects

suddenly unfold, but their green pigments (chlorophyll), the leaf veins, the sprouts and twigs are delayed in their maturation. Thus, these young leaves droop limply, their colour being either pale yellowish, reddish or even purple (Figs. 5.9 and 5.10). These false colours are also *mimicry*, the fresh young leaves feigning old, wilted—or poisonous—foliage to keep insects away.

5.5 The Soil of the Old-Growth Forest

Despite the unbelievably large variety of plant species and the lush vegetation growing everywhere in the rainforest, the primeval forest soil is extremely poor in nutrients. Even Alexander von Humboldt was deceived by the proliferating and "lavish" vegetation. He thought the soil was extremely fertile. In fact, the nutrients have been washed out of the soil over millions of years. What remains is a soil containing iron and aluminium, which is hostile to roots. This climax soil retains its limited fertile, balanced structure on its surface only as long as it is permanently moist—for example in the constantly shaded primeval forest soil. If it dries out after deforestation or slash-and-burn, only hard, strong reddish-brown *latsole* (*from* Latin *later* = *brick and*

Fig. 5.10 For a perfect illusion even false colours (here purple) are used

solum = soil) or *oxisole* (oxidized, i.e. used-up soil) or red loam soils remain. In other places, for example at black water rivers and lakes, you can find light bleaching soils *(podzol)*. All these soils are infertile, pure leaching soils (Fig. 5.11a). The reddish soils consist—as already explained—mainly of ferruginous and aluminous clinging soil ($Al^2 0^3$ and $Fe^2 0^3$), which is toxic to plants. It also lacks nearly all the clay minerals that enrich our native soils. When the thin humus layer disappears, for example, during road work, red earths harden into a stable ochre-red structure, which are called laterite (Fig. 5.11a). These climax soils are also very acidic because the lime has been washed out. In the untouched primary forest the uppermost layer is a dark, fertile humus covering these leached soils. This layer is only a few centimetres thick (Fig. 5.11a, b).

In our climate only few plant species with low vegetation could exist under such soil conditions. But the Amazon rainforest exists on one of the poorest and most washed-out soils on earth! Why can such a lush diversity of plants with an enormous biomass develop here in spite of this? How can this paradox be explained?

The surprising answer is: the forest actually only grows *on* and hardly *from* the soil. The soil is used primarily only as a substrate and as a mechanical fixation for the trees, and hardly at all as a source of nutrients. The explanation for the apparent contradiction between lushness on the one hand and poor soil on the other lies in the short-circuited *nutrient cycle* of the forest. Due to the high humidity and warmth, an extraordinarily fast remineralisation of the foliage and other organic waste such as excrements and animal corpses takes

Fig. 5.11 (**a**) Under the thin humus layer extends the climax soil (latsole), which is forever infertile, (**b**) This ochre-red hard soil structure is called laterite. It is also forever infertile

place up to a soil depth of about 10 cm. During the rainy season, soil foliage is completely decomposed within a month, and a fallen smaller tree is mummified by bacteria, fungi, ants and termites in just 1 year. Only the mosaic of the many specialized members of this ecosystem can maintain this rapid cycle of nutrients. With other words: In this closed system the lack of nutrients is compensated by a rapid recycling of the available nutrients. However, this nutrient cycle does not have an efficiency of 100%. For example, the 20% or so of nutrients get lost and must be compensated for by the Saharan dust that constantly falls, the canopy leaching and the rotting and weathering that comes from the soil. One hectare of the Amazon rainforest produces 33.5 t of biomass (mostly leaves) per year. A North American beech forest, on the other hand, produces only 13 t/ha. This tropical forest therefore lives in an almost closed nutrient cycle in which the superficial and extraordinarily dense root system (root felt) acts as a perfect filter for the nutrients (you can hardly get into the primeval forest soil with a spade).

However, the large primary forest trees still need the help of bacteria, ants, termites and, above all, fungi in the ground so-called *mycorrhizae* (a branching filamentous network of *hyphae*). This enormous network of fungus hyphae and tree roots spans beneath the forest floor and can be compared to an "invisible intelligence" (Suzanne Simard, University of British Columbia) (Fig. 5.12). The nutrients in the leaf litter are processed via the fungal filaments, which coat all root tips, and delivered directly to the root felt of the trees as nutrient salt ions (such as nitrogen, potassium, phosphate). There is a kind of symbiosis between the two participants: the fungal mycelium supplies the tree with nutritive salts and water, the tree roots supply the *mycorrhiza* with carbohydrates. In addition, specific chemical messengers are exchanged. For example, when herbivorous insects attack a tree in large numbers, specific stress signals are sent out, which are passed on to other trees via the underground networks between the tree and the fungal filaments, leading to chemical defence reactions in the latter. The removal of a single large, centuries-old "mother tree" from the rainforest, for example by cutting a special noble wood (this is the "large star" in Fig. 5.12), results in the loss of many such connections; clear-cutting destroys the entire network that has developed over centuries.

In the Amazon rainforest, more than 80% of the tree species have corresponding fungal species, so that extrapolations assume up to one million

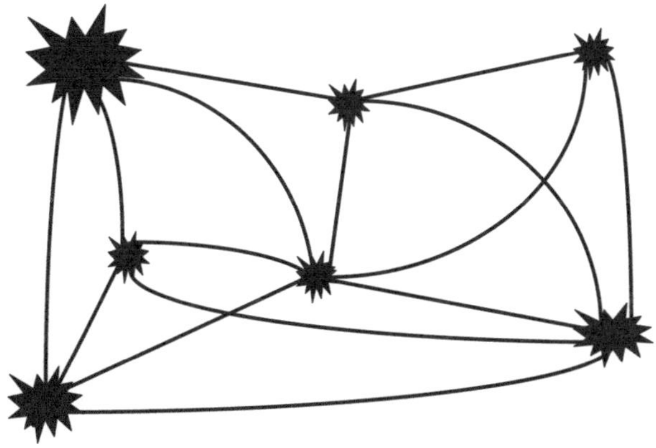

Fig. 5.12 Subterranean symbiotic network between the roots of rainforest trees and *mycorrhizae* (fungal filament network). Large star: old rainforest giant; small stars: younger trees

different soil fungi in the entire Amazon basin. The absolute amount of nutrients found in this climax stage, which is stable without human influence, is 80% equal. In this near equilibrium of all organisms, the same nutrients cycle through the jungle again and again—supplemented by the Saharan dust, soil weathering and canopy leaching. The dripping water from the canopy and from the trunk runoff of the trees consists of leaching from leaves and flowers as well as water flushing from excretions and is therefore chemically very rich and can be compared to a broth.

At this point, the great danger becomes clear when humans intervene in this wafer-thin balanced ecosystem. In the case of large-scale deforestation, the circulation of nutrient capital is cut off. After cutting and/or burning of the forest, the *mycorrhizal* network of innumerable soil fungi species is destroyed, 90% of the biomass in the form of bound nitrogen and 40% of all mineral nutrients in stock are lost immediately, and the fertility of the soil for planting is exhausted forever after only a few years. Man-made mineral fertilization would not help either, it is immediately washed out again and would have to be renewed every 14 days.

5.6 Top Floor of the Rainforest

The uppermost storey is formed by giant trees up to 60 m high, which tower above the entire leafy roof of the middle storey. I refer to these impressive tree giants as "cathedrals" of the rainforest in reference to the Cologne Cathedral in Germany, which also rises far above the sea of roofs (Fig. 5.13).

Why do these trees not grow into the sky? The answer is of a physical nature: due to the evaporation that constantly takes place via the leaf pores on warm, sunny days, there is a permanent water suction from the roots via the tube system in the wood to the leaves. Thus, a tree only 25 m high in the rainforest with a leaf area of about 1500 m^2 evaporates 400 L of water. The strong negative pressure in the leaves and the cohesion of the water molecules in the narrow tube system of the wood make this water ascent against gravity possible. The water molecules have different electrical charges and therefore literally "stick" together—according to the physical principle of "communicating tubes". This enables the water molecules to withstand high tensile stresses. At some point, however, every water pipe reaches its limits, contrary to gravity. If the water pressure is not sufficient, gas bubbles form in the water, interrupting the constant flow. If the water transport then falls below a critical limit, the tree stops growing in height. This limit is approximately at a tree

Fig. 5.13 The giant trees (here a kapok tree), which can reach heights of up to 60 m, tower above the closed canopy

height of 115 m (as in the case of the giant sequoia, *Sequoia sempervirens*, in California).

The giant trees in the Amazon rainforest are mostly kapok trees, also known as *cotton trees* (*Ceiba pentandra; Bombacaceae*), but also rubber trees *(Hevea brasiliensis; Euphorbiaceae)*, Brazil nut trees *(Bertholletia excelsa; Lecythidaceae)*, Colombian mahogany trees (*Cariniana pyriformis; Lecythidaceae*) and others. These solitary trees are exposed to special climatic conditions. For example, the temperature irradiation as well as the wind force are significantly higher for them than in the lower altitudes. These jungle giants are often several hundred years old and are home to hundreds of different species of epiphytes (Sect. 6.5) and lianas, as well as providing shelter for various animal species (such as night monkeys, bats, tree rats, snakes, agoutis and tarantulas) in their huge roots.

Interestingly, the kapok trees look quite different when they are still young. Then their slender trunk is green and full of sharp thorns. Only after several decades do the typical board roots develop (Fig. 5.14). When the fruit capsules are ripe, they open and release the approximately 100 black seeds, which are able to fly with the help of the long silk hairs attached to them. These hairs have a fine wax coating, making them water repellent. They are used as a high-quality fibre for life jackets, mattresses and pillows.

Fig. 5.14 The jungle giants develop board roots up to 9 m high, which stabilise the tree like a foundation

On one of those giant tree (namely *Luehea seemannii*), the number of beetle species was determined in a unique experiment. For this purpose, the US biologist Terry Erwin sprayed the entire foliage of the tree with the biodegradable *pyrethrum*, whereby all insects were stunned and fell to the ground. As a result of this experiment, 1200 different species of beetles were identified (for comparison, there are 5800 species of beetles on the total area of Germany). On another giant tree, 43 species of ants were identified, as many as the ones that live in the whole of Great Britain.

5.7 Root Types

Due to the special soil conditions (waterlogging and soil fertility decreases after only a few centimetres), only shallow-rooted trees grow in the lowland rainforest. One third of all trees usually develop enormous board roots. These reach up to 9 m up the trunk and only form wood on their upper side. In this way they stabilise the tree like a supporting pillar on the waterlogged, shallow soils. Where they enter the ground, there are short, turnip-shaped main roots, from which a huge number of fine lateral roots branch off at right angles. These form the root felt, which is connected to the *mycorrhiza* and supplies the tree with oxygen, nutrients and water.

In this oxygen-poor, constantly wet soil, there are two other characteristic root types. On the one hand, there are stilt roots, which branch off from the base of the trunk at a height of a few metres and anchor the tree firmly in the ground (as in the case of the walking palm *Socratea exorrhiza*, which grows to a height of 20 m). They even enable the tree to change its location over the course of a few years, for example when tree breakage nearby suddenly makes more incoming light available on one side. Then they increasingly form stilt roots in this direction, which shift the entire trunk in the new direction (see the arrow in Fig. 5.15). This is why it is nicknamed the "running palm". A number of fig species (*Moraceae*) also form roots branching from the main stem to support the tree in the soft soil. The third type of root is what are known as shoot-borne roots. In this case, the base of the trunk still above the ground—merges into thousands of roots with the thickness of a finger, which give the tree the necessary support; this is the case, for example, with the açai palm *Euterpe oleracea* (Fig. 5.16).

Fig. 5.15 *Socratea* palms ("running palms") develop tall stilt roots that increase their stability and enable the tree to move even in the direction of the arrow

Fig. 5.16 Shoot-born roots originating from the base of the stem, as in the case of açai palm

5.8 The Brazil Nut Tree

The Brazil nut tree (*Bertholletia excelsa; Lecythidaceae*) is a slender climax tree of the *terra firme*, up to 50 m high, whose crown projects above the middle floor (Fig. 5.17). It is native primarily in the Brazilian state of *Pará*. The diameter of its trunk can exceed 3 m. Alexander von Humboldt already knew about the importance of its fruits for humans. On his journey to the South American rainforest, he offered an ounce of gold to anyone who would bring him the large creamy-yellow blossoms of this tree. But he was there at the wrong time; no one could get him the blossoms. However, if the visitor wanders through the Amazon rainforest in February, he will find the almost intact flowers on the jungle floor under a Brazil nut tree, with only the ovary missing, remaining on the tree (Fig. 5.18). They flower high up at the top of the tree and for only a few hours in the morning before falling off. Thus the collector can check the very complex flower structure of this tree.

Furthermore, many Brazil nuts lie under the tree canopy between the decaying foliage. After only a few months, the shell becomes brittle and many seeds begin to germinate (Fig. 5.19a).

Here, too, a battle for life and death takes place. Ultimately, only one seedling will be able to grow into a tree, but until then the seedlings compete with each other for space and nutrients in the brittle shell. The last one that remains eventually feeds on its "siblings." 100–500 of these mighty capsule fruits are

Fig. 5.17 The slender Brazil nut tree grows up to 50 m high

Fig. 5.18 The flowers of the Brazil nut tree

produced by a mature tree within 22 months. They fall from the tree after this long ripening period without the lid opening. There are only a few animals in the rainforest that are able to break open the hard outer walls. These include the capuchin monkeys (Fig. 8.74), which beat the capsule on the ground until

Fig. 5.19 (a) Seedlings grow from the capsule fruit of the Brazil nut tree and compete with each other. Only one will survive. (b) Inside the almost spherical fruit of the Brazil nut lie up to 32 hard-shelled tasty seeds. However, the woody seed coat still has to be removed

Table 5.2 The components of the Brazil nut—The Brazil nut with "anti-aging" properties

100 g contain...	Proven effects
• Protein 13.0 g	• Antioxidant
• Sugar 2.8 g	• Analgesic
• Fat 66.0 g	• Blood pressure lowering
• Of which unsaturated fatty acids 25.5 g	• Anti-inflammatory
• Fibre content 8.3 g	• Immunostimulant
• Water 6.0 g	• Promotion of cell regeneration
• Na, Mg, Ph, K, Ca, Fe	• Aphrodisiac
• Vitamin E, C and other	
• Selenium 1.9 mg	

it breaks, and the agouti, which is able to open this hard fruit with its strong teeth (Sect. 9.8 and Fig. 9.13). Since it also buries the seeds in hiding places, it thus contributes to the spread of Brazil nut trees. The fact that the capsule fruits fall to the ground without the lid opening is good for the collector, as it allows him to pick up the fruits without losing the precious contents. The almost spherical fruit, which weighs up to 3 kg and is up to 20 cm in diameter, is enclosed in an extremely hard shell. Inside lie up to 32 triangular, hard-shelled seeds about 6 cm across (Fig. 5.19b). These must first be freed from their brown, woody seed coat before the tasty, ivory-coloured and hard flesh of the fruit can be accessed (Brazilian "Castanha de Pará"). With a fat content of 60%, it is very high in calories.

Other components and properties beneficial to humans are listed in Table 5.2. Incidentally, when a few years ago in the retiree's paradise of Florida an article in the daily newspaper described the "anti-aging" properties and aphrodisiac effect of Brazil nuts, all Brazil nuts were sold out in the supermarkets the following day.

In the Berlin Botanical Museum the visitor can find a herbarium specimen of a leaf of the Brazil nut tree, which was collected by Humboldt. A herbarium (from Latin *herba* = "herb") is a collection of preserved parts of a particular plant. It serves to describe the plant (for the first time), to compare it with other species and ultimately to classify it. Humboldt was the first human, who made an exact botanical description this tree. He named it after the French chemist Claude Louis Berthollet, who at that time had discovered the bleaching effect of chlorine. Humboldt chose *excelsa* (= excellent) as the species name.

In the following is described in detail how this tree species depends on other rainforest creatures in order to survive. Its way of life illustrates clearly that the primeval forest is more than the mere sum of its trees. There are many relationships between the inhabitants of the forest, which are absolutely necessary for their survival and which are not immediately apparent to us. If individual organisms of this network of relationships and food web are lost—for example through logging or burning—there is no survival for the remaining members of this alliance either.

The flowers of the Brazil nut tree, for example, are strangely narrowly constructed. The inside of the flower is helically wound and the nectar is located in the narrow centre of the spiral, where only certain insects with long proboscises can reach. The long-trunked females of a certain species of bright metallic or blue orchid bees *(Euglossini)* are able to do this. With their long proboscises they reach the tasty nectar of the flowers and pollinate the flowers in this process. However, these female orchid bees only mate with their male counterpart when the latter have perfumed themselves with the attractive scent of a very specific orchid species that only grows on Brazil nut trees or in their vicinity as epiphytes (Sect. 6.5). Without this orchid species, therefore, the orchid bees would not exist and without them there would be no Brazil nut trees. It is a mutual dependence that has evolved over the course of evolution. Because of this complicated triangular relationship, Brazil nut trees cannot be planted.

So these three totally different creatures are forever dependent on each other. If one species dies, so do the other two. Such relationships between different creatures, invisible to us humans, are numerous in the Amazon rainforest. We just don't know them all yet. Some more, for example the Queen Victoria's water lily (*Victoria amazonica*, Sect. 6.4.1) and the Arum lily (fruit of the devil or moco-moco), (*Montrichardia arborescens*, Sect. 6.3.2) are described in this book. If a thread were to be spun in the jungle between all these interdependent creatures, the result would be a gigantic web.

5.9 Epiphytes and Climbers

The uppermost floors of the tropical rainforest are colonized by a variety of *epiphytes (*literally, "top-down plant") (Fig. 5.20).

They prefer sunny locations and like to sit on weakened or already defoliated trees, as well as on trees along the riverbank. Epiphytic ferns also settle in the shaded area of the forest. Climbing plants, on the other hand, are also found in the lowest storey, which is exposed to twilight all day. Climbing plants and *epiphytes* are "squatters", but they do not damage the tree, so they are not parasites.

Epiphytes colonize branch forks and tree trunks, and depending on the light type, either the lower, low-light, temperated tiers (for example, nest ferns, *philodendrons*, bromeliads) or the all-day hot, sunlit exposed sites at the very top of the canopy (such as orchids). In both cases epiphytes have to withstand extreme environmental conditions. So most of them, such as bromeliads, have succulent—that is, thick-fleshed—narrow leaves, and many also form rosettes that collect rainwater. This tiny habitat is even home to certain tree-climbing frogs, which lay their few—usually no more than 10—eggs there, guard the hatching fry and, in the absence of mosquito larvae, even produce infertile

Fig. 5.20 Different *epiphyte*—species on a host tree. In the foreground, flowering representatives of the *Bromeliaceae* (pineapple family) can be seen

eggs *(abortive eggs)* to feed the larvae. If the leaf rosette dries out, the tadpoles are successively relocated to water-filled leaf rosettes (Sect. 9.1).

In addition, epiphytes have developed other strategies to get water and nutrients. For example, there are *bromeliad* species that form water-absorbing hairs on their roots, and some fern species collect falling leaves in their rosettes, which become humus. Orchids have aerial roots that are capable of absorbing rainwater and thickened shoot axes (so-called *bulbs*) that serve as storage organs for water and nutrient salts.

Climbing plants (lianas), on the other hand, have developed special tricks to get from the shady forest areas to the light-rich crowns. There are species whose seeds germinate in the forest floor and then grow upwards on the tree trunks with their thin and flexible, fast-growing shoot axes. There are different ways of doing this. There are root climbers. These twine their way up tree trunks. There are *rankers, left winder, and right winders*, which climb up their host (like the passionflower; see Sect. 6.6.2). There are spreading climbers, whose leaves and stems are transformed into hooked thorns, and adhesive rooted climbers. It is only when they have reached the treetops and thus the sunlight, do all these climbers first develop a kind of branched crown until they finally flower and fruit—often only after many years, sometimes even after 20 years. Climbers can live for many decades, their woody main stem can grow as thick as a thigh, and their stems often form strange free-hanging bends, loops and arches between trees. These are relics of the past, when the current bends grew up trunks that have long since ceased to exist.

A liana shoot often combines separate thin strands of tissue—like bundles of cables with thin wires, they are flexible and tough (Fig. 5.21) and can also reach 100 m in length. Lianas belong to many different plant families, such as *Passifloraceae*, *Curcubitaceae*, *Vitaceae*, *Convolvulaceae* or *Fabaceae*.

Furthermore, there are liana species, which germinate—in reverse—on their host tree, then first live *epiphytically* with direct access to sunlight. From there, they soon develop long, thin aerial roots that grow downwards, free-floating or clasping the host. These include above all the many representatives of the large plant families of the arum family (*Araceae*), the mulberry family (*Moraceae*) and the legumes (*Fabaceae*), such as the genus *Bauhinia*, called "monkey swing" because of its wavy ribbon-shaped shoots.

When these have reached the soil (or the water surface of a river), their (aerial) roots strengthen due to the increased nutrient and water supply and rapidly form many more secondary stems. Eventually—perhaps after 50 years—the size and areal extent of the original plant has increased 100 or even 1000 fold. Its crown now clearly exceeds that of its host. There are numerous fig and strangler fig species that practice this growth strategy.

Fig. 5.21 In an old liana shoot, thin strands of tissue, often separated from each other, are united—like bundles of cables with thin wires, so they are flexible and tough

However, there are also strangler fig species that remain on the original host tree for their entire life after they have come to germinate in the forest floor or on its branch fork. Their red or yellow fruits are very attractive to birds. They contain sugars, but they are also draining. The bird must relieve itself soon after eating them, and this it usually on a branch. So the seeds are deposited there, and it is not long before another strangler fig begins to germinate. The winner of the silent battle, even if it lasts many decades, is certain from the start: it is always the strangler fig. Over the years, it surrounds its host more and more with a tentacle-like bark and wood structure until this completely encloses the trunk of the host (Figs. 5.22 and 5.23) and literally suffocates it. This is the "strangulation point", the host tree is deprived of air and any possibility to grow, it dies. The strangler fig has taken its place, a free-standing tree up to 50 m high and with a crown of its own with a span of more than 40 m. This tree begins to flower and to grow and the cycle begins again.

Occasionally it happens that the stifled rotten original tree gradually decays, leaving a hollow cylinder, surrounded by the tentacled wood of the newly formed strangler fig (Fig. 5.24).

Fig. 5.22 A strangler fig at the beginning of its growth: tentacle-like, it begins to grow around its host

5.10 Summary: Structure of the Rainforest

The structure and composition of the Amazon rainforest from the ground to the heights are different from those of our forests in the northern hemisphere, although here as there all plants strive towards the light.

In summary, the primeval forest consists of three, not clearly distinct floors: near the ground, as already mentioned, there are only seedlings, 1–2 m long tree saplings ("oskars"), a few herbaceous plants and practically no shrubs because of the lack of light. On the shady ground lie also the fruits and seeds of the trees: the visitor finds few large and heavy seeds of the slow-growing hardwood climax species, which have not fallen far from their producer. They lie in waiting (which can last for a long time) as if in a seed bank. Especially near the riverbank, small seeds of light-demanding, shade-intolerant pioneer species such as the *Cecropia* are found in abundance (Sect. 6.1.1 and Fig. 6.1). Pioneer species, however, also live in the forest interior, but in small numbers. Their seeds are *dormant* (= "sleeping"), that is they are capable of long dormancy. They are therefore "prepared for all eventualities", because growth must happen very quickly when a sudden gap in the trees occurs. Even a short-term increase in white light in the form of wandering sunspots on the

Fig. 5.23 A strangler fig has already almost completely enclosed its host tree

forest floor has a germination-inducing effect on these light and temperature germinators, which is a competitive advantage.

The temperature at ground level varies by only one to two degrees Celsius in the course of a day, as hardly any sunlight reaches it (Fig. 5.25). The leaves of the plants are large, usually with a drip tip for rapid drainage of rainwater. About 14 species of birds have their habitat in this area of the rainforest. The few butterflies are mainly dark or metallic in colour. Flowers are rarely to be seen, but there are often remains of flowers on the ground that have fallen down from the canopy. Many visitors are therefore disappointed during a first hike through the forest, as their idea of the primeval forest is formed by painted pictures full of flowering plants, colourful birds and splendidly coloured butterflies. Only 5% of the rainforest is directly accessible to us

Fig. 5.24 After many decades, the battle is over: the strangler fig has taken the place of its host. The hollow cylinder marks the space of the former host tree

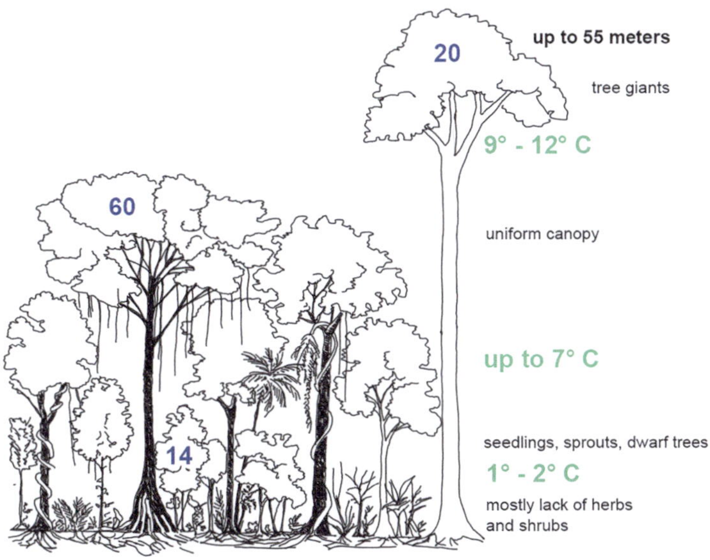

Fig. 5.25 Structure of the Amazon rainforest

humans, as our foresight is limited to 2 m in this world of trees. In fact, we humans move in the digestive tract of the forest. The fallen leaves and flowers, fruits and seeds, animal excrements and carcasses, loaded with nutrients, are "digested", or more precisely recycled, on the ground by soil organisms and the *mycorrhiza*. If you look more closely, you might notice a frog now and then, or different species of reptiles, for example lizards, anolis, Teiids (*Teiidae*) for example whiptails and junglerunners, skinks or even a snake, interesting species of ants (leafcutter ants, bull ants and termite burrows on the tree trunks (Sects. 9.10 and 9.11).

The second floor—the middle layer—is formed by a large number of trees that form more or less a uniform canopy. Here the temperature fluctuates around 7°C during the day and about 60 bird species have their habitat here (Fig. 5.25); they are heard but hardly seen. Two-thirds of all life here takes place in the treetops. On this floor and in the area of the tree giants, most of the trees as well as the epiphytes and lianas also blossom and fruit, but we can hardly see all this from the ground, perhaps with binoculars. The higher it goes up, the more colourful are flowers, butterflies, birds. These striking colours serve as camouflage, because they are hardly recognizable in the constant change of light and shadow, just like metallic colours (for example in the hummingbirds). This perfect camouflage acts like *somatolysis*: the body outline is dissolved in the mosaic of leaves in combination with light and shadow. The leaves of this floor are amazingly uniform in design. They are small, oval, coarse and usually finished off with a wax layer—all this serves to protect against evaporation during the high solar radiation.

The huge, age-old trees ("cathedrals") form the uppermost floor with their trunks and enormous crowns (Fig. 5.25). Here the temperature varies the most during the day, around 9–12 °C, and 20 bird species (see the blue number) have their habitat in the treetop. Sometimes the visitor sees the nests of the crested oropendola, also known as corn birds (*Psarocolius decumanus*), elaborately made from palm fibres and resembling weaver birds, which prefer to nest in tall kapok trees (Fig. 8.62b).

Our forests in the northern hemisphere, on the other hand, have three distinct layers, consisting of the herb layer (for example ferns, mosses and blueberry plants), the shrub layer (for example elderberry, common spindle or mountain-ash) and the tree layer (for example lime, beech or birch). There are very few lianas (for example common ivy) and epiphytes such as mistletoe are also extremely rare.

6

Flowering Plants in and Around Water

In the Amazon basin, the traveller is above all on

- the Amazon itself,
- its incredibly many, mostly white water, but also black water tributaries,
- the *Igarapés*, which are the narrow, natural connecting channels,
- the white-water and black-water lakes
- and in the rainy season also on the flooded *Várzeas* (river floodplains) and *Igapós* (floodplain forests)

on the way. On the banks of these waters, which receive sunlight throughout the day, are growing flowering trees, *epiphytes* (Sect. 6.5), lianas, shrubs and perennials, especially during the rainy season between February and May. All these plants of morphologically distinct groups can be seen from the canoe or dinghy. The tree species presented in the first part of this chapter are found in the forest interior or are occasionally planted in parks or as street trees in settlements along the Amazon. They are either deciduous (usually at the end of the rainy season) or evergreen. However, the latter does not mean that the leaves stay on the tree forever. No leaf lasts forever! Rather, each tree species has its own individual biorhythm as to when it sheds and renews its leaves. This can be the case with the "evergreen" species after 2 or 3 years or even after 5 or more years. Only on the main river—the Amazon—the visitor will normally not move close to the bank, so that flowering trees, *epiphytes* (Sect. 6.5 as well as Sect. 5.9) and climbers can hardly be seen, or only with binoculars. In addition, the riparian edge is subject to perpetual erosion, which inhibits the continuous development of plants. The second part of this chapter

L. Staeck, *Fascination Amazon River*, https://doi.org/10.1007/978-3-662-64452-2_6

introduces aquatic plants. These either float on the water surface of the predominantly current-free *várzeas* and white-water lakes or line the banks of weakly flowing *Igarapés* or small tributaries. During the rainy season, up to 20 different floating aquatic plants can develop and also flower in the shallow water.

6.1 Trees

6.1.1 Shield-Leaved Snake Wood Tree (*Cecropia peltata*)—Nettle Family (Urticaceae)

There is one group of trees that visitors always encounter during a trip along the Amazon: Pioneer woody plants with high light requirements, easily spotted on the banks of all white-water rivers and on the inner meander bends of the main stream, where sandbanks are deposited in the dry season. Most of them belong to the genus *Cecropia* that includes more than 60 different species. According to recent genetic studies, this genus has been placed in the nettle family (*Urticaceae*). They have predominantly slender, almost white trunks with soft wood and, as typical pioneer trees, they remain rather low at around ten to 20 m. The white trunks are an adaptation to the extreme location with up to 12 h of sunlight per day. The white bark reflects the sunlight, while dark trunks absorb the light. Their strikingly large silvery-green leaves, usually ten-pinnate on long petioles, are distantly reminiscent of our chestnut of the northern hemisphere (Fig. 6.1). The large leaf buds glow dark red on the tree and look like flower buds at first glance. The leaf circumference is usually nearly circular. These pioneer trees make no demands on special soil quality, and they are the first to germinate and grow up at lightning speed after a (natural) disaster such as fire or flood, or after the clearing of the virgin forest in the now open area: they grow 4 m or even more a year and reach a height of up to 10 m in 3 years. Even a sandbank newly formed after the end of the rainy season can thus be recolonised in a short time, as indicated above, so that in the following year, despite the onset of flooding, the rooted sandbank with its *cecropia* growth remains. During the rainy season, these trees stand in the water for months. They are truly the typical first colonizers of the Amazon rainforest. However, their lifespan is limited to about 20 years for most species, but during this time they prepare the ground for subsequent, more demanding tree species that already begin to grow under their crowns.

Fig. 6.1 The shield-leaved snake wood is a pioneer plant with a high light requirement

Cecropia are dioecious, which means that there are male and female trees. Both male and female inflorescences are composed of flower stalks with several spikes. The male ones are yellowish and consist of innumerable tiny anthers, the female ones are grey-green and look like sausages. The fertility of a single tree is—typical for pioneer trees—impressive. Extrapolations showed that a tree produces up to 900,000 tiny seeds per year, which are able to germinate for up to 10 years. The fruits are the favourite food of certain bats and in this way they ensure the mass dispersal of the seeds, especially along river banks. A number of fish species also eat the fruits that fall into the water and thus also contribute to the spread of the *cecropia*.

In German they are called indicatively "ant trees". Why? They live in a close community with the ant genus *Azteca*. More than 80% of the emerging young *cecropia*, which stand in the floodplains between April and October, are populated with colonies of this ant species, with numbers of individuals running into the millions. They are tiny reddish ants, barely 2 mm long. They are not called fire ants without reason. After I chopped off pieces of young *cecropia* shoots with a machete from the canoe to show that they were hollow inside and inhabited by these ants, hosts immediately swarmed out. They immediately pounced on me and my neighbors in the boat, biting aggressively. Each individual bite hurt fiercely for a short time like stings from pins, and with hundreds of them swarming on our bodies, it is easy to imagine that it takes a long time to catch the last biting ant. Figure 9.16 was accordingly produced with considerable physical effort.

Most of these *Cecropia* trees live in close symbiosis with these ants (Sect. 9.10). Climbing predators, such as leafcutter ants, are also detected, immediately attacked and driven away. The excretions of the millions of individuals also provide these trees with a valuable fertiliser. However, one predator is overlooked: the sloth, which apparently moves so slowly on the *cecropia* that the ants simply do not recognize it while eating its favorite juicy leaves.

Tip

When cruising along the riverbanks by boat, it's worth using binoculars to scan the tops of the trees for sloths, which—optimally camouflaged—look like a pile of wilted leaves and sit high up in the cecropia trees.

Incidentally, the *Caboclos* use the leaves of the shield-leaved snake wood (*Cecropia peltata*) for medicinal purposes. The leaf substances lower the blood sugar level and a tea made from the leaves is drunk against flu-like infections. For this, 100 g of leaves are brewed with 1 l of boiling water and the brew is drunk.

6.1.2 Gustavia (*Gustavia superba*)—(Lecythidaceae)

Another conspicuous tree species that likes to settle as understory tree or on such banks of white-water rivers that are seasonally flooded, is the Gustavia tree or heaven lotus. Its hand-sized, splendid flowers glow like candles in the green of the leaves from April onwards. This evergreen tree of 6–10 m in height, usually grows like a bush without a discernible top, was named after the Swedish King Gustav III (1746–1792) (Fig. 6.2). The species name *superba* means "sublime" and refers to the conspicuous, pleasantly fragrant flower. The mostly terminal flowers are in racemes, with usually only one or two flowers open while the rest are still in various budding stages of development. The creamy white spoon-shaped petals, covered with a pink tinge at the beginning of flowering, are particularly attractive. In the centre of the flower is a mass of equally attractive pale yellow stamens arranged in tightly packed rings, their lemon-yellow anthers curving inwards. This striking arrangement looks like a hoop. The conspicuous cup-shaped woody fruits, about eight inches across, are closed by a flat lid. The family name refers to the shape of

Fig. 6.2 The *Gustavia tree* has delicately fragrant, splendid flowers

the fruit (*lecythus* = jug). The *Gustavia tree* is related to the Brazil nut tree, both fruits have a similar shape. In the Brazil nut, however, the lid remains closed even after the seeds have ripened so that the fruits fall to the ground with all their seeds when they ripen and can thus be easily collected. The lid of the fruit is blown off when the seeds ripen, revealing numerous tiny, shiny black seeds with a yellow seed coat, called *aril*.

6.1.3 *Eschweilera ovata*: (Lecythidaceae)

This evergreen tree with a spreading crown grows to a height of 18 m and prefers to grow on the banks of white water rivers that are flooded during the rainy season. Thus, when it matures, its seeds fall into the water and contribute by this way to spread this tree species. During its flowering period in May, the decorative flowers that have fallen into the water can still be seen several kilometres downstream of its habitat, still intact except for the receptacle with the ovary, floating in the water like small yellow sails. This flower dropping is typical of many members of this family, such as the Brazil nut and paradise nut, which drop their nearly intact flowers. This makes it easy to study and photograph the flower structure, which is otherwise impossible as the flowers sit high up in the treetops. The fragrant flowers usually sit in a short raceme. The flower structure of all members of this family is complicated and not easily recognizable. In *Eschweilera*, the numerous, lemon-yellow stamens sit crowded in a narrow ring at the center of the six large cream-colored petals

Fig. 6.3 The attractive flowers of the *Eschweilera tree* are captivating because of their numerous, ring-shaped, lemon-yellow stamens, which are located in the centre of the cream-coloured petals

(Fig. 6.3). Over this ring is a yellow cap on one side, partially covering the stamens. Pollinating insects must crawl under this hood to reach the pollen, pollinating the flower in this process. The roundish capsule fruits, which are 10 cm in size, remain on the tree after ripening—in contrast to the Brazil nut tree, for example. However, their cap is blown off, releasing the two seeds—often into the water. The seeds, which are covered with a cream-coloured seed coat (*aril*), are eaten by rodents such as squirrels, or spiny and tree-rats and fish.

This tree was named after the German botanist Franz Gerhard Eschweiler (1796–1831), the species name *ovata* means "ovoid" and refers to the shape of the fruit.

6.1.4 *Campsiandra comosa*: Legume Family (Fabaceae)

This medium-sized tree, which barely reaches a height of 12 m, also prefers to grow along white-water river courses. It is most recognisable when the light to chocolate-brown, woody, slightly curved pods, up to 25 cm in size, hang in large quantities from the branches in March/April (Fig. 6.4). When these

Fig. 6.4 With the onset of the rainy season, innumerable 25 cm long woody pods hang from the *Campsiandra tree*

pretty fruits are allowed to dry, the fruit shells break open in a spiral to reveal up to ten, nearly round, shiny chestnut-brown, hard-shelled seeds that are considerably larger than a two-euro coin (about 4 cm in diameter). The local population uses the ground seeds as an addition to pastries. The bark extract is said to stop diarrhea, and residents in Spanish-speaking regions use the bark juice, under the name *Huacapurana*, to fight tick-borne borelliosis.

The pretty, spike-shaped inflorescences, visible from October onwards, contain numerous small, five-petalled flowers with white petals, which—before they open—are coloured blood-red on their outer sides. Particularly striking are numerous long, white-red stamens bearing a dark red anther (Fig. 6.5).

The genus name is composed of *campsis* (= to bend) and *aner* (= male) and refers to the shape of the stamens in some species. The species name *comosa* means "crested" and thus refers to terminal inflorescence.

6.1.5 *Parkia discolor*: Legume Family (Fabaceae)

This evergreen, up to 15 m tall tree with a dense crown is a sun-loving, fast-growing pioneer tree that prefers to stand on the flooded banks of the *Várzeas*, but also likes to grow on the edge of the black-water Igapós—without any special soil requirements. Sandy, nutrient-poor soils are completely sufficient for this tree species. It probably lives in symbiosis with soil bacteria, which

Fig. 6.5 The flowers of the *Campsiandra tree*, which are blood-red before flowering, are densely packed together in an inflorescence

supply it with the necessary nitrogen. Its pretty leaves are pinnate, consisting of double pinnate leaflets with up to 50 individual leaflets.

The *Parkia tree* is easily identifiable when it flowers between August and November because its inflorescences are unique, strange and unmistakable (Fig. 6.6)! They look like the fruiting bodies of a fungus and are scarlet in color. The long-stalked inflorescences hang from branches, making them eminently accessible to their nocturnal pollinators: nocturnal bats. During the day, bees come to collect nectar. Each inflorescence has an interesting "division of labour":

The topmost sterile flowers with the long, hair-shaped, bright red, flower-like stamens (the "mushroom cap") serve to attract pollinators. Below this is a narrow zone of small, sterile nectar-producing flowers. The lower section (the "mushroom stalk") is covered with many tiny fertile flowers. The up to 15 cm long, reddish-brown pods (Fig. 6.7) contain seeds, which are also on the menu of *caboclos*.

The Scottish explorer Mungo Park (1771–1805) is the eponym. The species name *discolor* means "different coloured" and probably refers to the different red colouring of the inflorescence.

Fig. 6.6 The mushroom-like inflorescence of the *Parkia tree* is unique

6.1.6 Guiana chestnut (*Pachira aquatica*: Mallow Family (Malvaceae)

Other Names: Provision Tree, Malabar Chestnut, Monguba (Brazil)

This evergreen deciduous wetland tree with a height of up to 14 m and a broad spreading crown is a sun-loving pioneer tree that likes to stand in swamps for long periods. It is therefore more common in the floodplain meadows of the *Várzeas*. Its stalked leaves are compound and palmate. Between September and December single extremely decorative flowers appear at the end of the branch, up to 15 cm long, resembling large shaving brushes (Fig. 6.8). Between the four long, lanceolate, first creamy yellow, later dark red petals, sit many long white stamens, red at the tip, with red, later yellow elongated anthers. Between them the long, red pistil rises. These flowers are

Fig. 6.7 The fruits of the *Parkia tree* grow into 15 cm long, red-brown pods

pollinated by bats. The imposing woody brown fruits, up to 30 cm long, harbour numerous nuts (Fig. 6.8), which are eaten raw or roasted by the *caboclos*.

The Guiana chestnut or the similar species *Pachira glabra* is also a popular houseplant because of its decorative leaves, which can be kept indoors all year round. Cut leaf stalks root easily when placed in a glass of water.

The genus name *Pachira* is a proper name of the Indians in Guiana, where this tree was first identified. The species name *aquatica* means "in the water" and refers to the preferred location of this species.

In the past, this tree species was classified as a member of the kapoc family (Bombacaceae).

Fig. 6.8 The decorative flowers of the Guiana chestnut resemble a shaving brush. Its fruits, up to 30 cm long, harbour numerous edible nuts

6.1.7 Munguba Tree (*Pseudobombax munguba*): Mallow Family (Malvaceae)

Other Name: Brazilian silk-cotton-tree

This fast-growing, sun-loving pioneer tree grows up to 30 m tall. It is dependent on damp sites and survives standing in water for months without any problems. It is therefore mainly found in flooded meadows of the Várzeas. Its imposing reddish-brown fruits, which remain on the tree at the end of the rainy season in May while the broad leaves gradually fall off, are striking (Fig. 6.9). The seeds are eaten by monkeys, bats and fish. In addition, dispersal occurs through fruits falling into the water, which are readily eaten by fish.

Just as impressive as the fruits are the apple-sized flowers. They appear after the leaves fall in the dry season. With the beginning dusk they start to blossom and glow like candles in the moonlight of the rising night. The flowers—being functional for a single night only—are pollinated by bats. They seem to consist only of creamy-white stamens (Fig. 6.10), crowded close together in prodigious profusion around the sturdy white pistil with dome-shaped stigma projecting from their centre. The coarse, whitish, lanceolate petals quickly wither at the base of the flower.

Fig. 6.9 The imposing reddish-brown fruits of the Munguba tree still hang on the tree when the leaves have already fallen off

Fig. 6.10 The creamy white flowers of the Munguba tree seem to consist only of stamens

This tree species used to be classified in the plant family Bombacaceae, but today it is new classified in the subfamily Bombacaceae. It is related to the baobabs of Madagascar, East Africa and Northwest Brazil. The term *Bombax is* derived from the Greek *bombyx*, meaning silk: the seeds of the kapok tree (*ceiba*) exhibit silk-like flight organs on their seeds. The tree described here is related to the kapok tree, but it is a "false *bombax*" (= *pseudobombax*). The species name *munguba* is a proper name from the *Tupi language*.

6.1.8 Clitoria Tree (*Clitoria fairchildiana*): Legume Family (Fabaceae)

Other Name: Sombreiro tree

This plant is more of a large shrub than a tree in an appearance on the banks of blackwater bodies and more rarely on whitewater. This is due to the short trunk and the dense crown spreading above it. This is why the *Clitoria* has the Brazilian nickname "Sombreiro" (meaning like a hat that gives shade). As a sun-loving, fast-growing and flood-tolerant plant, it is a pioneer woody plant that likes to establish itself in deforested areas. It lives in symbiosis with nitrogen bacteria, which process atmospheric nitrogen in its roots and make it available to the cells. As in many other legume plants, the roots contain *ruthenone*. This significant phytochemical belongs to the *isoflavonoids* or *fucocumarins*, which have anti-inflammatory properties. The native population uses the root extract in fishing to stun fish.

After flowering in April/May, the foliage leaves begin to fall and the bean-like, brown fruits ripen. In June they burst open explosively and fling the purple seeds far away. The decorative flowers, about five inches long, sit in terminal racemes. They are pale purple with purple narrow longitudinal stripes and are slightly fragrant. It owes its German and international name *Clitoria* (Fig. 6.11) to its striking resemblance to the female vulva, including the *clitoris*. It is a popular street and park tree in Brazil. The species name *fairchildiana* honours the US botanist David Fairchild (1869–1959).

6.1.9 Ice-Cream Bean (*Inga edulis*): Legume Family (Fabaceae)

Other Name: Inga tree

This tree species used to be classified in the mimosa family (Mimosaceae), but today it belongs to the subfamily of this plant family and is assigned to the legume family. This evergreen tree, which can reach a height of up to 15 m, loves sunlight and is tolerant of flooding. It likes to grow in marshy

Fig. 6.11 Because of the shape of its flowers, this tree is called the "shameless" or *clitoria*

Fig. 6.12 (**a**) The filigree flowers of the *Inga tree*, (**b**) The slender fruits, up to 50 cm long, are the reason for the popular name "monkey's tail"

floodplains along white-water rivers. It is considered a pioneer tree that is often found in secondary forests. Members of the genus *Inga can* be easily identified by the conspicuous abundance of their mostly white, but also yellowish, often curly stamens (so-called powder puff flowers). The pinnate leaves are composed of four to six pairs of large, oval leaflets. Terminal inflorescences appear between August and November. The stalkless, slightly fragrant single flower consists of a rather inconspicuous, whitish sepal tube from which a large number of elongated, white stamens, tubularly fused at the base, rise (Fig. 6.12a). The equally narrow, white inconspicuous pistil rises from their centre. The fruits, which ripen between March and May, are imposing. They are legumes up to 50 cm long and 1–3 cm wide, often twisted and black when ripe (hence the nickname "monkey's tail"; Fig. 6.12b). Breaking open the fruit reveals the creamy-white flesh, in which 8–15 shiny black seeds are enclosed. This fleshy creamy mass tastes pleasantly sweet, almost vanilla-like, and is eaten directly from the bean, the seeds being spat out. Amazonian markets frequently offer these fruits. The pulp is made into sweet desserts such as puddings and ice cream. Of the more than 250 *species of Inga*, many different endemic species grow in the Amazon region whose fruits are also edible, for example, the yellow-flowered *Inga pilosula*. The syrup-like sap is taken as a tea against bronchitis or used to heal wounds. The wood is used to make furniture.

The genus name *Inga* is a proper name in the *Tupi language*, the species name *edulis* translates as "edible".

6.1.10 Marimari Tree (*Cassia leiandra*): Legume Family (Fabaceae)

This sun-loving pioneer species also likes to colonize seasonally flooded black and white water banks as well as secondary forests near lagoons and canals. It does not mind flooding for several months. Marimari is a smaller, deciduous tree about 10 m tall. Its leaflets consist of 9–12 pairs of leaflets 4–7 cm long. The terminal, pendulous inflorescence, 15–20 cm long, may contain more than 30 individual flowers and unfolds from September to December. The flowers are a decorative lemon yellow and are composed of five sepals and five petals. The four stamens and narrow pistil are also bright yellow (Fig. 6.13). The fruits, up to 50 cm long, 3 cm wide, roundish and slightly curved with distinct indentations, ripen between March and June. They do not open at full maturity. The fruits are harvested around February when they begin to turn from green to yellow-brown. After tearing open the fruit walls, you can see that the individual round seeds, separated by sturdy septa, are surrounded by a ring of green pulp (Fig. 6.14). You grab a seed enclosed in the pulp with your fingers and suck the seed off, then spit it out. This pulp tastes pleasantly sweet and sour—like the children's lemon drops. This is why these fruits are popular in regional markets.

Worldwide, the genus *Cassia*, which is restricted to the tropics, comprises more than 700 different species, of which a number—always yellow-flowered—are also endemic to Amazonia.

Fig. 6.13 The Marimari tree has large yellow flowers with conspicuous stamens

Fig. 6.14 The green tasty pulp surrounds the ring-shaped seeds of the Marimari tree

6.1.11 Garlic Pear (*Crateva tapia*): Caper Family (Capparaceae)

This evergreen, sun-loving small tree, often remains shrubby, reaching a growth height of only about 8 m even in good conditions. It likes to stand on the banks and is flood-tolerant. Its fruits, when ripe, fall into the water between January and June, where they are eaten by fish and thus further dispersed. The terminal inflorescences, which appear between August and September, have flowers in racemes. The outer flowers sit on longer stems than the inner ones, so that all the flowers reach a common level. The decorative individual flowers consist of four white petals arranged in opposite directions. Up to 20 long reddish stamens protrude from the interior of the flowers and soon descend (Fig. 6.15a). The somewhat thicker pale pink pistil sits in

Fig. 6.15 (**a**) Inflorescence of the garlic pear, (**b**) The round, slightly garlic-scented fruits of the garlic pear are readily eaten by fish

their centre. Hard even when ripe, the greenish-grey, mandarin-sized fruits are edible and have a sweetish taste (Fig. 6.15b). When cut open, they give off a slightly garlicky odour.

The genus *Crateva* occurs pantropically, i.e. worldwide, and was formerly classified as a member of the Capperaceae.

6.1.12 Many-Leaved Acacia (*Acacia polyphylla*): Legume Family (Fabaceae)

Surprisingly, there are acacias in the Amazon area, which are actually at home in Australia. This is because South America together with Africa, Australia, Madagascar and Antarctica used to form the large southern continent of *Gondwana*. Today, the vast majority of the well over 1000 species of Acacia grow in Australia. The species described here is a sun-loving pioneer tree that likes to grow along riverbanks. There it catches the eye of passing visitors when it is in full bloom between December and April. As a pioneer tree, it also likes to colonise secondary forests. Every few years it also sheds its leaves when its fruit ripens between August and September.

This acacia is a thorny tree with a round, dense crown that reaches up to 20 m in height. Its decorative leaflets, up to 26 cm long, are divided into 10–16 leaflets, each 6–8 cm long. The inflorescence is a terminal panicle of white flower heads. The flat, brown pods, about 6 cm long, spring open at maturity and release their seeds into the water.

The genus name comes from the Greek *akakia*, which means "thorn". Acacias used to be counted among the mimosa family, but today they only belong to the subfamily Mimosoideae. The species name "polyphylla" means "with many leaves".

6.1.13 Morich Palm (*Mauritia flexuosa*): Palm Family (Arecaceae)

Other Name: Buruti palm (brazilian)

This palm species is unmistakable among the more than 300 palm species in Brazil due to its stately growth form and its fruits. It is a decorative character plant, especially in the lower reaches of the Amazon and its tributaries. It is sun-loving and likes to grow in swamps and periodically flooded riverbanks. It develops a mighty trunk up to 30 m high, with a diameter of up to 80 cm.

Its crown is also particularly spreading and hemispherical, as the living leaf fronds usually grow upwards or horizontally. The leaf fronds are also very wide, up to 4.5 m, and its petiole is also extremely long, up to 9 m. The enormous inflorescence consists of numerous, curtain-like, cream-coloured lateral branches about 60 cm long, on which the small flowers are densely packed. The enormous reddish-brown fruit clusters ripen between December and June and are already clearly visible from a distance (Fig. 6.16). The seeds are dispersed by the current and by many different animals such as agoutis, tapirs, peccaries, monkeys and fish, which are particularly fond of these fruits. The numerous fruits, up to 5000 in 1 year, 8 cm long, spherical and very decorative, are covered with large scales (Fig. 6.16). This palm species serves the local population as a food and useful plant. Alexander von Humboldt, therefore, called it "the tree of life" in his writings: "so many things are linked to the existence of a single palm. (…) Thus, at the lowest level of human mental formation, we find the existence of an entire tribe of peoples tied to almost a single tree" (A. v. Humboldt: Ansichten der Natur. Berlin 1807, p. 33).

Fig. 6.16 The striking, large, red-brown fruit clusters hang from the mighty Morich palm. The edible, decorative fruits are covered with large scales, which can be peeled off with the fingernail after watering

When the fruit is soaked for 24 h, the leathery skin can be easily peeled off with a fingernail. Then the lemon-yellow flesh becomes visible, which houses a hard nut. Rich in vitamins C and A, this pulp is eaten raw, made into flour or fermented into alcohol. An orange-red oil can also be extracted from the pulp, which contains three times more vitamin A than carrots. It also contains *oleic acid* with monounsaturated fatty acids and *carotenoids*. It is used to make sweet dishes. When the inflorescence is cut off before flowering, a sweet sap flows from the cuttings, which is fermented and eventually yields a good-tasting *buruti wine*. The buds of the inflorescence are also used as a vegetable. A sago-like starch is obtained from the stem tissue. The leaf fibers provide ropes and hammocks, and the pith of the leaf stalks is used to weave mats. Finally, the thick larvae of a palm weevil can be "harvested" as *suri grubs* once they have introduced their eggs into the trunk tissue (Fig. 3.2). Ultimately, the thick trunk provides wood for house construction and pier installations. Humboldt was therefore quite right in his statement.

The genus name *Mauritia* honours the German Prince Moritz of Nassau (1567–1665). The species name *flexuosa* means "bent, curved" and refers to the growth habit of the fronds.

6.1.14 Buçu Palm (*Manicaria saccifera*): Palm Family (Arecaceae)

Other Names: Ubuçu (brazilian), Manicaria

This palm species is also so distinctive in appearance that it is unmistakable. It is immediately noticeable when driving along the banks of white-water rivers. Its leaflets are gigantic and measure up to 10 m, making them among the largest leaves in the plant kingdom (Fig. 6.17). After dying, they remain on the tree even longer. This species is also more often multi-stemmed, although the trunks themselves tend to be short, at 3–5 m tall. The base of the trunk is enlarged with many roots. It grows with a preference on sandy soils as well as on flooded meadows (*várzeas*) and is sun-loving. Its fruits hang in clusters on the tree all year round. They look like small coconuts. The local population uses the milk of unripe fruits as a medicine against coughs. The trunk is used as timber, the leaves for roofing.

The genus name *Manicaria* means "like a sleeve" and refers to the shape of the bracts of the inflorescence, the species name *saccifera* means "bag-bearing" and also refers to the shape of these *bracts*.

Fig. 6.17 The buçu palm has huge leaflets up to 10 m long

Yet another palm is easily recognizable on the banks of the white-water rivers: the slender açai palm with its small mop of leaflets. This is described in Chap. 4 with Figs. 4.31 and 4.32.

6.2 Shrubs

6.2.1 Mickey Mouse Bush (*Ouratea lucens*): (Ochnaceae)

Other Names: Carnival Bush, Chupeta de Macaco (brazilian for "Monkey Soother")

This plant is a richly branched, ornamental shrub up to about 3 m high, which grows in the floodplains of blackwater lakes and rivers and stands almost completely in the water for up to 7 months in the rainy season without dying. The oval leaves, about ten inches long, are entire, often strikingly glossy. The small flowers are golden yellow, with the five sepals and petals each on the gap. There are numerous yellow stamens in the centre. The numerous pistils are fused together and form a common style. At the base of the flower, however, they are isolated from each other, resulting in a funny arrangement when the fruit ripens (see the name).

The fruits surprise with their flower-like shape and colour. The initially green-olive, later black, elongated-oval one to two drupes sit separately on a bright red, fleshy swollen receptacle. When two black fruits sit on the roundish-red receptacle, it looks like an oval Mickey Mouse head with two black ears (Fig. 6.18), which gave the shrub its name. Sometimes there is even an additional drawing of Mickey Mouse on the swollen former flower base, even the eyes are not missing. The genus name *Ochna* means "wild pear tree" and refers to the shape of the leaves. The species name *lucens* means "shiny" and refers to the surface of the foliage leaves.

A rather similar flower is shown in Fig. 6.19, but this one comes from a tree in the blackwater flood forest of Amazonia called *Heisteria spruceana*, so named in honour of the German botanist Lorenz Heister (1683–1758). Brazilians also like to refer to these flowers as "monkey soother". They belong to the Olacaceae family, and the local population also calls this tree *Muruchi*.

Fig. 6.18 The two black fruits of the Mickey Mouse bush look like the ears of the comic strip character of the same name. The additional drawing on the red flower base completes the impression of a Mickey Mouse figure

Fig. 6.19 The flowers of the *Heisteria bush* look similar to the Mickey Mouse bush

6.2.2 Guarana (*Paullinia cupana*): (Sapindaceae)

This plant is a woody evergreen shrub that can climb to a height of about 12 m, even under favorable conditions—for example, on the riverbank. From cut thin branches rubbery latex milk is flowing. This bush can only be recognised among the many plants on the banks when its decorative fruits, about 2–3 cm in size, ripen between February and March. Then they turn orange-red to red and partially open. This reveals two (or three) jet-black seeds, 12 mm long, surrounded at their base by a white, fleshy *aril* (seed coat). The striking color of the fruit attracts birds. These eat the seed coat with seeds and thus spread this plant.

This arrangement of black and white is very attractive and looks like an eye. Numerous legends of the *Guaraní people* also tie in with this. The name of the plant is also borrowed from this Indian ethnic group "Guarana" means "fruit like the eyes of the people". While traveling along the banks of blackwater rivers and lakes, the visitor is more likely to see this wild form of guaraná, such as the related species *Paullinia pinnata* (Fig. 6.20a), rather than the larger plantation fruits offered in the market only in December. The wild fruits look somewhat different from the commercially sold *Paullinia cupana* (Fig. 6.20b). The small, rather inconspicuous, whitish, unisexual flowers are pentate and

Fig. 6.20 (**a**) The wild form of the guaraná shrub *Paullinia pinnata*, (**b**) The fruits of the "real" guaraná look like eyes

bloom between November and December. The seeds contain caffeine (3–6% or 40–90 mg per gram dry matter) and are used therapeutically by the local population. The ingredients have a stimulating and performance-enhancing effect and reduce the feeling of hunger. Today, guaraná fruits are a dietary supplement, the plants are cultivated in plantations near the Amazon. The peeled and dried seeds are ground into a light brown powder, dissolved in water, sweetened and drunk as a refreshing drink or energy drink. Without added sugar, the powder tastes tart and bitter. The tannins—contained in the guarana powder—are slightly astringent and delay the effect of the alkaloid caffeine. As a side effect, sleep disorders, tachycardia (rapid heartbeat), headache and muscle pain are described.

I found another Paullinia species with a decorative yellow seed coat on the bank of a white water stream.

The first description of this plant species was made by Karl Kunth, a botany professor, who evaluated the plant samples brought by Humboldt in 1801. The genus name *Paullinia* honours the German physician Simon Pauli the Younger (1603–1680).

6.2.3 Golden Trumpet (*Allamanda cathartica*): Dogbane Family (Apocynaceae)

Other names: Allamanda, Liane à Lait ("Milk Vine"), common trumpetvine.

This sun-loving shrub, up to 6 m tall, likes to grow on the banks of white water rivers. There it also likes to climb as a climbing shrub. The shrub is easily noticed by visitors travelling along the riverbank, as it always flowers very profusely and has a pleasant sweet fragrance. As it is very fond of flowering, it is often planted in Brazil (and now even worldwide in tropical regions) as a ground cover in hotel sites and parks. The glossy leaves are equipped with a drip tip. The almost spherical, large capsule fruits are covered with spines up to 2 cm long. Its decorative sunny yellow, waxy and funnel-shaped large flowers are in terminal umbels, the throat appears slightly reddish (Fig. 6.21a).

The family name is composed of *apo* (= away) and *kynos* (= dog), which should mean: Keep these plants "away from the dog" because their milky-looking sap is poisonous. The genus name refers to the Swiss botanist Frederick Allamand. He sent seeds from Surinam to Carl von Linné in 1770 who, in turn, gave the genus name. The species name *cathartica* aptly means "laxative" or "intestinal cleansing" and thus refers to the poisonous parts of the plant (see also the name of this plant family).

Fig. 6.21 (**a**) The golden trumpet has large flowers resembling the shape of the brass instrument, (**b**) The sister of the golden trumpet, *Allamanda violacea*, grows as a delicate-limbed climber in floodplain meadows

The caterpillars of the frangipani sphynx moth feed on the leaves of the golden trumpet. They are almost immune to the latex (see Sect. 8.7). Despite its poisonous effect, the indigenous population sometimes uses the milky sap of the leaves as a laxative. In the cities, cultivated dark red varieties can also be seen. In the *várzeas*, between February and May, the related species *Allamanda violacea* grows, a decorative, delicate-limbed climber conspicuous by its purple-red flowers (Fig. 6.21b). In the northern hemisphere, the golden trumpet is now available in nurseries. It is easy to root via green cuttings.

6.2.4 Blue Potato Bush (*Solanum spec.*): Nightshade Family (Solanaceae)

Other Names: Gentian Tree, Potato Tree

The nightshade family includes more than 3000 species worldwide. Mostly they are shrubs of different growth height, whereby the largest can reach up to 6 m and sometimes—as with the two species described here—appear like small-growing trees. They are sun-loving, often standing on the edge of white-water floodplains (*várzeas*), and do not mind flooding for several months. Especially the species described here are among the first to settle when the primary forest has been destroyed. The leaves are often large and lobed, especially when old. A distinctive feature is the formation of soft hairs on all parts of the plant when young, and the formation of spines on the leaves, the shoot axis and the calyx. In *Solanum crinitum* the large, rich purple flowers, very similar to those of the potato, are composed of five-petalled, star-shaped, green sepals and also five-petalled, largely fused, curled petals (Fig. 6.22a). The flowers, initially purple, rapidly fade and eventually become almost white.

Fig. 6.22 (**a**) Blue potato shrub *Solanum crinitum* has curled petals, (**b**) Because of its bizarre udder-shaped form, the species *Solanum mammosum* is often planted in gardens

In the centre is the prominent yellow stamen cone; the stamens are pointed upward. The berry fruits, first greenish, later greyish-green and the size of mirabelles, are often edible, but not in the species described here. The fruit is often still enclosed by the sepals, especially at the beginning. Tubers as in the potato, as we know them, do not exist. In the related species *Solanum subinerme* the five deep purple petals are narrow, smooth and deep. In this plant, the fruits are about the size of a cherry and turn from yellow to red when ripe.

The genus *Solanum* is distributed mostly in South America, especially Brazil, where all habitats from dry savannas to floodplains are colonized. However, this genus is also found in south-east Asia, Africa and even in tropical Australia. Because of their elegant, richly flowering splendour and beautiful flower colour, some species are now also widespread pantropically as ornamental shrubs. The genus name comes from *solamen*, which translates as "comfort or soothing agent" and refers to the pain-relieving effect of some species. The species name *crinitum* translates as "hairy." The genus *Solanum* includes both the potato and the tomato. There are numerous genetic hybrids between the two groups in South America. In Brazil, also along the Amazon, I have discovered numerous *Solanum* species, some of which have tasty red fruits in the size of currant to a cherry. Some species are even of local importance, such as Lulo (*Solanum quitoense*), whose 3 cm large, yellow or reddish fruits occur from Peru to Amazonia. Because of its bizarre, soother- or udder-shaped, yellow, also orange, non-edible fruits, the species *Solanum mammosum* (="strong-breasted") is popular in the Amazonian lowlands (Fig. 6.22b). It is mainly the *solanines* and bitter-tasting *glycosides* that exert their healing, for example, laxative, effects in small doses, but lethal effects in larger doses. All nightshade species are undosed in principle toxic in their green parts—including the potato and the tomato.

6.2.5 Monkey's Brush (*Combretum aubletii*): White Mangrove Family (Combretaceae)

Other Names: Spanish Flag, Monkey Comb, Flor De Cepillos (Spanish = "Brush Flower"), Escova-De-Macaco-Alaranjada (Portuguese = "Orange Monkey Brush")

Very fast-growing, sun-loving, twining shrub up to about 15 m high, woody at the base, climbing, with dark green, coarse, lobed leaves up to 10 cm in size: the longleaf always grows on strongly sunlit white-water river banks. Even from a distance, passing visitors notice the orange-red brush-like flowers in the green clutter of the many riparian plants. The spikey inflorescence with hundreds of individual flowers stands on an axis up to 15 cm long, like the brushes of a toothbrush. At its base the single flower is tubular. The ovary (which is not visible) is also located there. Particularly striking are the very numerous stamens, up to 3 cm long (as in all members of this plant family), which attach to the upper edge of the bright red small flower cup. The anthers are pink. The stamens are first lemon yellow and then turn bright orange-red as they age (Fig. 6.23), creating the bicolour of the inflorescence (similar to

Fig. 6.23 Because of the numerous, upright stamens, the Amazon residents call this plant "monkey's brush". In the related species, *Combretum cacoucia*, the flowers hang in clusters on the branch

the flag of Spain). The plant signals with the change in the flower's colour to its pollinators, especially butterflies and hummingbirds, when the nectar supply is particularly abundant, namely when the stamens turn yellow. The tamarin monkeys and the white-fronted capuchin monkeys also like to feed on its nectar, making both monkey species important pollinators of this plant (Sect. 8.6.4). The calyx does not change colour. The 2 cm large fruits are winged.

Another *Combretium species* can also be found along the riverbanks, especially in secondary forests: *Combretum cacoucia.* This species is also a climbing shrub—or it grows like a climber up to more than 10 m high on suitable supporting plants. The terminal, tufted, pendulous inflorescence is large at about 20 cm and harbours numerous decorative individual flowers about 2 cm in size. These consist of a rich red, fused, five-celled calyx from which ten stamens, also red and up to 2.5 cm long, protrude (Fig. 6.23, upper right). These are typical hummingbird flowers.

The scientific genus name *Combretum* derives from the Roman name of an unrelated creeper. The species name *aubletii* refers to the French apothecary Jean Aubletiana (1720–1778).

6.2.6 Odontadenia *hypoglauca*: Dogbane Family (Apocynaceae)

Other Names: Agadenia hypoglauca dipladenia (in nurseries)

This grateful semi-shrub, which also likes to climb like a vine, is sun-loving and likes to grow in white-water floodplains. Its almost palm-sized, funnel-shaped flowers feature an unusual color scheme: An apricot color pattern stands out in the centre of the propeller-like twisted, five lemon-colored petals, which looks like a second, five-zip flower (Fig. 6.24). As members of the dogbane family, the branches contain a white, poisonous milky sap. The leaves appear blue-green on their underside. The elongated-elliptical-shaped fruits are also special. They always grow in pairs (i.e. two by two) and are huge at about 20 cm long and 5 cm wide. In good nurseries other representatives of the genus *Odontadenia*, for example very decorative, red-flowered species, are offered under the name *Dipladenia.*

The genus name *Odontadenia* is composed of the two Greek terms *odonthos* (= tooth) and *adén* (= gland) and thus refers to five toothed small glands under the ovary. The species name *hypoglauca* translates as "blue-green".

Fig. 6.24 The almost palm-sized flowers of *Odontadenia hypoglauca* have a colour pattern in their centre that looks like a second flower

6.3 Perennials

6.3.1 Hanging Lobster Claw (*Heliconia rostrata*): (Heliconiaceae)

Other Names: Lobster or Crab Claw, parrot beak (French: Bec de Perroquet)

This up to 3-m tall, non-woody perennial with banana-like leaves is sun-loving and grows with a preference on the banks of white-water rivers, also on light-flooded forest edges and clearings, along paths and in secondary forests. Its coiled leaf sheaths—the lowest part of the leaf—form a pseudostem. The leaves are up to 1 m long and resemble banana leaves. In the species described here, the powerful inflorescence hangs out of the mock stem on a long red stalk. The inconspicuous, white or white-greenish tubular flowers are hardly visible, as they are enclosed in red-yellow *bracts*. These large *bracts* hang down like garlands or "lobster claws" (Fig. 6.25). There are several endemic species of a heliconia in the Amazon lowlands. There is one species, where the inflorescences stand like flames on a shoot (*Heliconia caribaea*, see Staeck 2021a, p. 49 f.) that grows through between the leaves of the mock stem. In another species, the *bracts* are not red but golden yellow (*Heliconia psittacorum; Golden Torch*, see Staeck 2021a). The fruit capsules of heliconias contain three large, cobalt-blue seeds, making them particularly attractive to birds.

Fig. 6.25 The large red-yellow bracts of the hanging lobster claw hang down like garlands

Originally native to the Amazon lowlands, these heliconias are now widespread throughout the tropics, often running wild, but also in parks.

The genus name refers to the mountain *Helicon* in Boeotia/Greece, the seat of the Muses, the species name *rostrata* means "beaked", as the yellow-green tip on the red bracts of its inflorescence resembles a beak. There are about 150 different species of the genus *Heliconia* with a wide variety of growth habits, flower shapes and flower colours. With their attractive flower colours, they are typical "bird flowers". Thus they are visited by hummingbirds in South America. Often, the bawl-shaped bracts are filled with rainwater. These mini-aquariums are often a colourful aquatic biotope: small crustaceans, mosquito larvae, beetle larvae, worms, protozoa, even tadpoles of frogs with close interactions of eating and being eaten live in these confined spaces. Because the colourful and robust, and thus extravagant, inflorescences last for weeks until they fade, they have been exported to Europe and the United States for years, where they are offered for sale in decorative flower arrangements. Jamaica and Hawaii are the main exporters. In the regions of origin, the leaves are popularly used as tablecloths or packing material for wrapping food. By the way: The genus *heliconia* was formerly included in the family *Musaceae*, which includes the bananas.

6.3.2 Arum Lily (*Montrichardia arborescens*): Arum Family (Araceae)

This large perennial, up to over 3 m tall, is sun-loving and grows only on shorelines and in marshes of the white water. The residents of the Amazon lowland call this plant moco-moco. The further the visitors travel downstream from *Manaus* to the lower reaches of the Amazon, the more common it becomes. Large populations are particularly conspicuous on the banks of the *Brevis canals*. Their habitat extends up to the brackish water and mangrove regions. At low water, the stout rhizomes can be seen, from which shoot beginnings emerge vegetatively from the mud, much like asparagus tips. The mature, bamboo-like mock stems are impressive and can reach 25 cm in diameter (Fig. 6.26). Their large, arrow-shaped, hard leaves are among the favorite foods of the skunk birds (hoatzins) and the manatees (Sect. 8.5.11). As in all members of the arum family, the inflorescence consists of sheath-like bracts (*spatha*)—in this species 20 cm tall and creamy white—that surrounds the fleshy flowering bulb half open. This bulb is covered with innumerable, tiny, also cream-white flowers, from top to bottom with a fertile male, infertile male and fertile female flowers (Fig. 6.27).

Fig. 6.26 The mighty pseudo-stems of the arum lily push out of the bank mud like asparagus tips

Fig. 6.27 The inflorescence of the arum lily consists of white bracts and a fleshy, cream-coloured flower bulb. In order to attract the beetles required for pollination, the temperature of the flower bulb increases to over 40 degrees Celsius (see the red colour in the picture!)

The pollination mechanism in the arum lily is unusual and similar to that in the Queen Victoria's water lily *Victoria amazonica* (Sect. 6.4.1): with the onset of dusk, the intrinsic temperature of the flower bulb increases considerably after the barren male flowers in the middle section of the bulb have bloomed, reaching over 40 °C, so that there is a large temperature gradient with the current ambient temperature of about 26–28 °C (Fig. 6.27). Stored reserve fats are consumed in the process. This enormous energy expenditure corresponds to that of warm-blooded animals, for example, hummingbirds. This warming intensifies the release of the aromatic, sweet scents of the flowering bulb into the evening air. These two parameters, together with the white colour of the flowers at nightfall, act as irresistible sexual signals for beetles of the genus *Cyclocephala*. These beetles invade the flower bulb and feed on the pollen, stigma secretions and nectar of the sterile flowers. In doing so, they pollinate the female flowers that are ready for conception at this time with the pollen from another flower bulb that they are already carrying on their backs. To exclude self-pollination, male and female flower organs mature at different times. In parallel, the beetles copulate among themselves. These beetles look similar to our cockchafer ("maybug"), only slightly smaller; both belong to the large family of *Scarabaeidae*.

If you cut off a flowering bulb in the afternoon—around October they are in flower—you can experience the phenomena described for yourself at dusk, around half past six. Both the rise in temperature of the flowering bulbs as well as the perfume-like scent can be clearly registered. The fruits, about half the size of a pineapple and also with scaly skin, are edible when cooked and are called *Fruit of the Devil* because of their irresistible taste.

The family name is derived from Aaron, who is reported in the Old Testament to have held a staff in his hands, which later sprouted—probably in allusion to the staff-shaped fruit bulb in this plant family. The genus name honours Gabriel de Montrichard, who was living in Trinidad when the name was given in 1854. The species name *arborescens* means "tree-like" and refers to the enormous growth form of this plant species.

6.3.3 Dormilones (*Mimosa pudica*): Mimosa Family (Mimosaceae)

Other Names: Meaning-plant, touch-me-not, shame flower

This prostrate herbaceous plant, up to 40 cm high, has spiny stems. The spherical flowers, about 1 cm in size, are pink to light purple (Fig. 6.28). The leaves are pinnate and arise from petioles. The inconspicuous fruits, 4 mm wide and 2 cm long hang together in a cluster. Originally from the arid regions of Brazil, the seeds of the *Dormilones* have spread to all tropical regions via shoe soles and stockings, so that this plant is now considered a tropical "weed" and can also be found in Amazonia on sunny river banks and in cities on ruined areas (for example, former ruins).

The genus and family name goes back to the term *mimos*, which translated means "actor", the species name *pudicus* means "shameful". Those who are hypersensitive as humans have to put up with being called mimosas. When suddenly darkened, when approached by a source of heat, when shaken, but especially when touched most slightly—even in rain—the leaflets of this plant

Fig. 6.28 The flowers of the Dormilones are pink to light purple

fold up, and if irritation continues, the petioles also fold down. For this purpose, hinge-like "joints" are located on the petioles of the leaflets and the individual leaflets. The touch stimulus is transmitted—depending on its strength—up to 5 cm per second. The plant does this by transporting potassium ions into the cells at the base of the leaf, which trigger a change in the cell internal pressure (*turgor*). After about 10–15 min, the leaves straighten up again. Therefore, this plant is a prime example in biological education of the ability of plants to perform movements. Folding the leaves and snuggling up to the spiny stem may provide some protection from predators.

There is even a yellow-flowered sister of the *Dormilones*, the water mimosa (*Neptunia natans*), which lives in the *Várzeas* as a floating leaf plant, but otherwise has all the characteristics described above. It is also known under the synonym *Neptunia oleracea*.

6.4 Aquatic Plants

A special experience for every visitor to the Amazon basin is a trip by canoe or Zodiac into the *várzeas*, which often extend for dozens of kilometres into the hinterland behind the banks of the main river at high tide (Sect. 2.2 and Figs. 2.7 and 2.8). Along shallow sections of the Amazon's wall-like riparian margin, nutrient-laden water masses from the main river constantly flow into the

Fig. 6.29 Up to 20 different aquatic plants grow in the floodplain meadows

shallow hinterland as the rainy season progresses. In addition to the impressive bird and fish life in this habitat, it is above all the great variety of attractive, mostly flowering aquatic plants that share the quiet bays of the nutrient-rich white-water lakes between the floating grass meadows. Even the untrained eye will easily come across as many as 20 different aquatic plants growing side by side in the stagnant waters of a *várzea* (Fig. 6.29). The plant species described below can be found from early April at the latest, with many of them in flower. Unless otherwise indicated, all the plants presented grow in white water. In black water, on the other hand, none of the mentioned abundance of plants can be seen due to the acidic water quality (pH 3.8–4.9) and nutrient deficiency. Apart from the genus *Cabomba* (hairy moth family, Cabombacaceae), which is represented in black water by a few—submerged—species, there are only a few other aquatic plants (for example, the bladderworts (*Utricularia*)).

6.4.1 Queen Victoria's Water Lily (*Victoria amazonica*): Water Lily Family (Nymphaeaceaea)

This imposing, largest aquatic plant in the world grows only in the *várzeas*, the stagnant, nutrient-rich floodplain meadows on both sides of the Amazon

Fig. 6.30 In the calm and nutrient-rich floodplain meadows, Queen Victoria's water lily can develop magnificently

(Fig. 6.30), and more rarely in *igarapés*, the natural connecting channels between the Amazon River loops.

It is the only water lily that blooms *twice*! But first things first. With the onset of the rainy season in January, the first leaf buds begin to grow from the rhizomes, the fleshy roots, left behind in the dry mud. The first leaves open while still underwater (with an arrow-shaped leaf form), and from the fourth leaf onwards the water surface is reached. From then on, the leaf area becomes larger and larger, eventually reaching up to 3 m in diameter, making it the largest leaves of all aquatic plants. They are circular and have a margin of about 4 cm, which prevents other aquatic plants from growing onto the leaf surface. There are some gaps in this margin so that the large amounts of rain can drain away quickly. In addition, there are tiny holes in the leaf blade that also serve to quickly remove rainwater, as any drop of water left behind can act as a burning glass. There are extensive air chambers between the leaf veins to help the heavy leaf float. On the reddish underside of the leaf are multiple branching stiffening ridges that provide stability for the huge leaf; the leaf veins branch whenever the distance to the neighboring vein exceeds a certain size. There are additional tracheae in these ribs. The leaf can thus easily carry a weight of up to 50 kg. My little niece Anja was able to sit on the leaf plate in the Berlin Botanical Garden without any problems. The underside of the

Fig. 6.31 The underside of the leaves of Queen Victoria's water lily is armed with centimetre-long spines to keep predators away

leaf is also armed with centimetre-long spines to deter predators (Fig. 6.31). Not till the 20th leaf appears—in April/May—the first flower buds are formed.

The long-stalked flower buds, almost the size of a child's head, open almost abruptly at the onset of dusk on the first day with snow-white petals (Fig. 6.32)—the observer can almost follow the opening with the naked eye—smelling delightfully (a scent mixture of pineapple and butter cookies). The stigmas of the carpels are ready for conception. The yellow stamens, which are present in large numbers (up to 300!), are now barren. Four different beetle species of the genus *Cyclocephala* are attracted by the scent and the increasingly warming flower interior: the temperature is up to 11 °C higher than the ambient temperature. In particular, it is the cockchafer-like *Cyclocephala hardyi* and the smaller species *Cyclocephala verticalis*, which is the size of European June beetles (summer chafer), that crawl into the flower interior and feed on the yellow, protein-rich feeding bodies (Fig. 6.33). As they feed, they pollinate the stigma of the pistil with pollen from other flowers carried on their backs. Towards midnight the flower closes and the beetles are enclosed. By the forenoon of the second day, the petals have turned pink, with the pedicel distinctly stretched above the surface of the water. In the afternoon, at about 3 o'clock, the petals open a second time, now in a dark reddish-purple (Fig. 6.34), but without exposing the inner cavity.

Fig. 6.32 At dusk on day 1, the Queen Victoria's water lily blooms snow-white

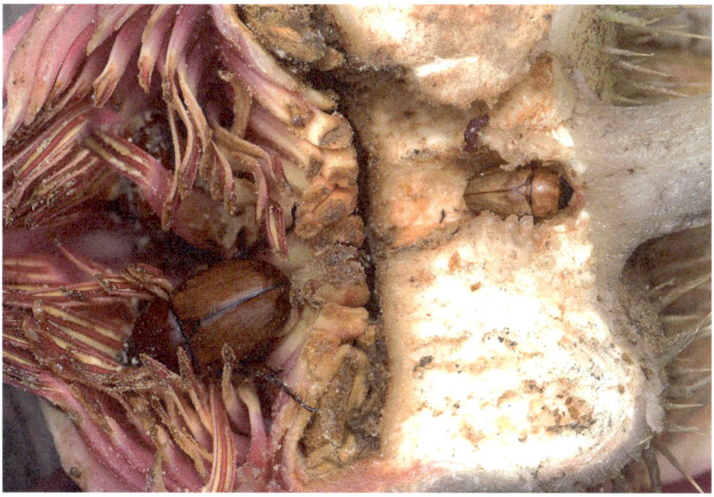

Fig. 6.33 Different beetle species are attracted by the scent, high temperatures and the candle-light colour of the flower and crawl into the flower interior

At dusk, the flower finally releases the beetles. The pollen has now ripened. The beetles that have been trapped overnight are released and—pollinated with ripe pollen—fly away all at once in search of white, fragrant, warm and nourishing water lily flowers that are blooming for the first time. Once there, they leave the pollen on the stigma, which is already ready to conceive, and use it to pollinate the ovary. The co-evolution of the two species of organisms

Fig. 6.34 At the end of day 2, the flower has changed to a dark red-purple color

makes them both dependent on each other: The beetle needs the water lily for food, and the water lily needs the beetle for cross-pollination. On the second night, the flower sinks below the surface of the water, where either the seeds mature or the flower rots. The fruits are usually not visible because they ripen under the surface of the water. They are fist-sized, spiny and contain a large number of lentil-shaped seeds, which are eaten raw or cooked by the locals.

This magnificent plant can now be found in many botanical gardens around the world, even in temperate latitudes—but there in greenhouses, where it also regularly comes into flower.

This plant was discovered by the German botanist Eduard Poeppig in 1832. Another botanist, Richard Schomburgh, suggested naming it *Victoria regia* in honour of Queen Victoria, who was reigning in England at the time. However, since Poeppig was the first describer, he also had the right to name it. Thus it was renamed *Victoria amazonica*. The net-like framework of the leaf veins on the underside of the leaf served the British horticulturist and architect Joseph Paxton as a model for the imposing *Cristal Palace* greenhouse at the first World's Fair in London in 1851, which was made only of glass and cast-iron struts. This was a significant building built according to the new science of *bionics*, which takes nature as a model.

6.4.2 Common Water Hyacinth (*Eichhornia crassipes*): Pontederia Family (Pontederiaceae)

Other Names: Thick-stemmed water hyacinth, Eichhornia, purple curse, blue devil

In the nutrient-rich, white-water floodplain meadows, this plant species seems to explode. When there is a large supply of nutrients, they rise up to half a metre above the water surface, otherwise—especially when there is a large current—they remain lower. On the main stream and its tributaries, floating leaf carpets of *Eichhornia*, grass and reed plants, sometimes more than 100 m^2 in size, float downstream everywhere during the rainy season when the current is strong (Fig. 2.4). On these, an instrumental orchestra of different frog species and cicadas has sometimes made themselves at home. When such an island of floating leaves rushes past the visitor, a loud, polyphonic concert of the most varied pitches is temporarily heard.

The oval, water-repellent leaf surfaces with blunt tips form a rosette on longer petioles. The leaf stalks are thickened at their base like blisters or sponges, so that air is stored. This gives the plant a fantastic buoyancy so that it can easily reach the most remote corners of the Amazon with the current. The richly branched main roots bear many full-bearded, bluish-black lateral roots that collect sediment from the white water. Various smaller daughter plants are often attached to the mother plant on side shoots.

From April/May, the decorative light purple flowers, which resemble hyacinths (but are not), appear on a spike-like inflorescence about 15 cm long. The inner upper flower lobes show a bright yellow spot (Fig. 6.35).

This plant species is originally only native to the tropical waters around the Amazon and has natural enemies there, for example, beetle and moth species. But because of its pretty flower, it has now spread to all continents with tropical climates via the intermediate stage Florida. It is now one of the ten worst "weeds" on our planet. Outside the Amazon region, it clogs the waters so that boats can no longer get through. Since in many places it covers the surface of the water without any gaps, underwater life is endangered. The waters then often overturn and the decreasing flow velocity of the waters allows the Anopheles mosquito to spread easily as a malaria vector.

Alexander von Humboldt discovered this striking aquatic plant during his South American expedition in 1801 in the Colombian *Cauca*. A good two decades later, the German botanist Carl Sigismund Kunth gave it the scientific genus name *Eichhornia*, with which he wanted to honour the Prussian

Fig. 6.35 Characteristic of the common water hyacinth is the bright yellow spot on the upper floral lobes

Minister of Culture Johann Albert Friedrich von Eichhorn (1779–1856). The species name *crassipes* means "pot-bellied" or "thick-stemmed". The family name refers to the Italian botanist Giulio Pontedera.

Their multiplication rate is not linear—as is normally the case—but exponential (Chap. 10). This means that a mother plant can produce an incredible number of 50,000 daughter plants within 6 months. Thus, in non-native areas where natural enemies are lacking, it quickly becomes a "biological run amok". That is why it is also called "purple curse", "water plague" or "blue devil" in the introduced regions. In the countries of the northern hemisphere, it has become a popular plant for the garden pond in summer. In this climate, however, it cannot cause any damage, as it is not hardy. Nevertheless, its importation into the USA and EU has been officially banned since several years. Its relative *Eichhornia azurea* (= sky blue) occurs in the areas of *Eichhornia crassipes*, although not as explosively. However, its petioles are not thickened, the edges of its inner petals are fringed and the flower throat is dark purple.

Another member of the Pontederia family, whose flower has similarities with *Eichhornia*, is also found in calm *Várzea waters:* Round-leaved pickerelweed (*Pontederia rotundifolia*, synonym *Reussia rotundifolia*). It roots in muddy soil or floods in calm várzea regions. The succulent, i.e. fleshy, leaves are ovate-round, arrow-shaped notched and rise with their stem clearly above the water surface. The single flower is pale pinkish-purple with a yellow spot

Fig. 6.36 Round-leaved tropical pickerelweed (*Pontederia rotundifolia*) flowers in a spikey, pale pink inflorescence only 4 cm across

in the upper third-like Eichhornia-but the single flower itself is barely a centimeter across and sits along with more than 50 other flowers on the spike-like, densely flowered inflorescence, which is only about 4 cm long (Fig. 6.36). The three dark stamens with mostly transverse black anthers are conspicuous. With this inflorescence, pickerelweed is not to be confused with *Eichhornia*. All parts of the plant are edible raw or cooked and are used by the locals as an ingredient in salads and soups.

6.4.3 Water Cabbage (*Pistia stratiotes*) Arum Family (Araceae)

Other Name: Shell flower

This aquatic plant is ubiquitous in all *várzeas*, white-water lakes and rivers in the Amazon basin. On the Amazon itself, many "islands" of water lettuce and of common water hyacinth float downstream. The decorative bright green

Fig. 6.37 The light green leaf rosette is characteristic of water cabbage. The tiny inflorescence consists of a white bract and an equally tiny yellowish flower bulb

leaves of the shell flower are striking (Fig. 6.37). They sit practically on the surface of the water and are arranged rosette-like. The broadly spatulate leaf blade is weakly notched at its tip. At night the leaves close. If you fish a plant out of the water, you can see the long-bearded, blue-black roots that filter nutrients out of the water. Normally this plant always floats on top of the water, but when the water level drops it also roots firmly in the soil. If there is a good supply of nutrients, for example in human sewage as well as in stagnant water, it can reach a diameter of over 40 cm, otherwise, it remains smaller. It reproduces strongly vegetatively via runners and can thus quickly colonize stagnant water surfaces without gaps, similar to *Eichhornia*. If it enters black water with the current, it does not survive but dies after a while. Its inflorescence is only 10 mm small, and can only be found on larger plants. To do this, you need to bend the petals apart so that the centre of the leaf rosette is exposed. There—as in all members of the arum family— sits the spathe, a petal-like cream-coloured bract forming a sheath to enclose the flower bulb. In the species described here, however, the bract is tiny and densely hairy on the outside. Opposite the even smaller, yellowish flower bulb sits with fertile and sterile individual flowers (Fig. 6.37). Particularly striking are the yellowish male flowers arranged in a ring at the tip of the small bulb.

Water cabbage looks appetizingly fresh like crisp lettuce, but it contains oxalic acid crystals (*calcium oxalate*) *that* are not recommended for human consumption in high doses even when cooked, but it is edible for pigs. The *caboclos* apply the sap of the plant to remove warts.

Water cabbage is a popular, easy to maintain aquarium plant in the United States and Europe, which can be purchased in nurseries. Its origin is not entirely clear. Some authors claim that it is pantropical, i.e. also at home in tropical Africa and Asia. I, on the other hand, think that it is native to the Amazon lowlands, partly because of its co-evolution with some insect species. Anyway, in fact, in many tropical African and Asian countries, it is an invasive, that is, invading, highly proliferating species that pushes back native plants. Therefore, in these countries (but not in the Amazon lowlands) it is considered a "dangerous weed" that has to be controlled.

The genus name *Pistia* is Greek and means "water", the species name *stratiotes*, also Greek, means "soldier".

6.4.4 Floating Heart (*Nymphoides humboldtiana*): (Menyanthaceae)

Other Names: Humboldt's sea pot, Nymphoides Indica

This decorative, perennial aquatic plant is reminiscent of water lilies with its leaves floating on the water—hence its generic name *Nymphoides* (= similar to water lilies). It grows mainly in nutrient-rich, stagnant waters such as lakes or marshes, where it can also form large clumps. It forms large numbers of runners at the rootless nodes of its stems. Its leaves are similar to those of water lilies but are considerably smaller at about 5 cm. The flowers stand in groups at the nodes above the water surface. They are radially symmetrical, the white petals fading to yellow at the base of the leaf and have a strongly fringed margin. The flower organs are pentate. The dark brown pistil has two to three pale, thick-fleshed lobes (Fig. 6.38a).

This genus occurs worldwide, also in temperate latitudes. The species described here is dedicated to Alexander von Humboldt (*humboldtiana*).

Fig. 6.38 (**a**) The leaves of the floating heart look like a mini water lily, but its decorative flowers appear much more delicate and smaller, (**b**) The mosaic flower forms pretty geometric patterns on the water surface, (**c**) The clearly visible white breathing roots next to the flower of the floating Ludwigia give this plant a fantastic buoyancy. A few water ferns (Carolina azolla) float next to the flower

6.4.5 Ludwigs

Ludwigiae (*Ludwigia sedioides*): (Onagraceae)

Other Name: Ludwigia

This perennial, attractive aquatic plant is conspicuous for its mosaic-like leaves. It prefers wet, marshy, nutrient-rich sites. It is occasionally seen in floodplain meadows (*várzeas*), where it forms pretty geometric patterns (Fig. 6.38b). When exposed to strong sunlight, they are often coloured reddish by anthocyanins to protect them from harmful UV rays. This plant species is also capable of forming submerged leaves. Its breathing roots are filamentous. The decorative radial-symmetrical, yellow flowers are only seen sporadically.

This Ludwigia is a popular aquarium plant. It is named in honour of the German physician and botanist Christian Ludwig (1709–1773). The species name *sedioides* refers to the stonecrop (*Sedum*) and means "Sedum-like", as the leaves of the *Ludwigia* resemble it.

Floating ludwigia (*Ludwigia helminthorrhiza*): *Evening primrose Family (Onagraceae)*

It is a perennial plant that floats in the stagnant water of the *Várzeas* or creeps in the marsh, rooting at all nodes. Its striking white respiratory roots, 1 cm thick and up to 2 cm long, allow it the necessary gas exchange in the oxygen-poor mud and also a fantastic buoyancy so that it always floats on the water surface. Its stems are up to 1 m long and often branched. Thus it often grows from a floating plant island into open water areas. Its nearly circular, entire-edged leaves, about 2 cm in size, are often reddish in color. The individual flowers have white petals that are yellow at their base (Fig. 6.38c).

Alexander von Humboldt collected and herbarised this plant together with his friend Aimé Jacques Alexandre Bondpland. Thus it could later be identified and named by the German botanist Carl Sigismund Kunth. The species name *helminthorrhiza* means "worm-shaped root" and refers to the shape of the stems.

6.4.6 Leafy Bladderwort (*Utricularia foliosa*): Bladderwort Family (Lentibulariaceae)

The perennial, herbaceous leafy bladderwort is an unusual plant. It has no real roots, and leaves and shoots can hardly be distinguished. Above all, it is a *carnivorous* plant that likes to grow where white and black water mix or in pure black water lakes or swamps that are very acidic, i.e. have pH values around or below 5. There it forms larger populations. Due to the lack of nutrients, it is dependent on an additional supply of protein by catching micro-organisms living in the water with a sophisticated catching apparatus. Its submerged, fine lilac leaves form an often matted web. Within this tangle of fine leaflets and shoots are numerous trap-like catch bladders, 2–3 mm in size (Fig. 6.39). These function according to the suction trap principle: a negative pressure is initially built up within the trap bladder. The trap bladder is closed with a flap that only opens inwards. On the outside, some bristles are attached. The algae-like leaflets now attract prey, for example, water fleas, threadworms or rotifers. When a bristle is touched, the door opens and the prey is sucked in by the negative pressure. The opening and closing of the door takes place within two milliseconds. Internal glands digest the prey

Fig. 6.39 The leafy bladderwort is a carnivorous plant. It catches its prey, e.g. water fleas, with numerous catch bladders

protein, which benefits the plant. When canoeing or zodiacing across a blackwater lake, the only way to recognize the leafy bladderwort is by its yellow flowers, which rise up to 20 cm above the surface of the water at the tip of an erect stem with air chambers in a contrast worth seeing to the tea colored water (Fig. 6.39).

The genus name *Utricularia* means "small tube" and refers to the shape of the trap bladders. The species name *foliosa* means "leafy" and refers to the immense number of very finely pinnate leaflets.

6.4.7 Aquatic Ferns

Surprisingly, a number of ferns, i.e. flowerless plants that reproduce via spores, also coccur in the biocoenosis of aquatic plants in floodplain meadows (*várzeas*). All of the aquatic ferns presented below are highly prized by aquarium enthusiasts in Europe and the United States.

6.4.8 Floating Antler Fern (*Ceratopteris pteridoides*): (Pteridaceae)

Other Name: Eagle fern-like horn fern

This very decorative 1-year fern prefers calm water, as we often find in the *Várzeas*. There it can reach 30 cm in diameter, otherwise, it remains smaller in shape and size. When the water level decreases, it can also take root in muddy soil. Its light green floating leaves, which may be submerged at first, are undivided when young, later broadly lobed with a broad triangular outline. The finely pinnate leaves protruding from the water are attached to the broadened petiole. On older leaves, when living conditions are good, many young plants develop from brood buds in the notches between the lobes (as in "mother-of-thousands", *Bryophyllum*). The roots are hair-shaped, fine and arise directly from the leaf base. The petioles widen strongly towards the leaf blade (hence its name antler fern). They consist of spongy, air-filled tissue and serve as floating bodies. On adult specimens, spore-bearing leaves (spore fronds) develop. These project steeply upwards and differ considerably from the sterile leaves. They are multipinnate and have narrow segments. The edge of these leaf segments is curled downwards and hides the spore capsules.

The local population is said to use the leaves as a vegetable, but literature indicates that they contain carcinogenic ingredients.

The family name derives from *ptéris* (= fern). The genus name is composed of *kéras* (= antler) and *ptéris* (= fern), the species name *pteridoides* means "fern-like" and refers to the antler-like shape of the spore fronds.

6.4.9 Eared moss (*Salvinia auriculata*): (Salviniaceae)

It is impossible to imagine Amazonia without this small floating fern. It can be found floating on the water surface in all stagnant or slowly flowing nutrient-rich (white-water) bodies of water. As a very invasive plant it is now found throughout the tropics and subtropics worldwide. I have even found it in the rice fields of Madagascar or Japan. It always consists of two light green, roundish leaves about 2 cm in size that rest on the water surface (Fig. 6.40).

Fig. 6.40 The eared moss consists of only two ear-shaped floating leaves. On the upper side of the leaves there are countless whisk-shaped papillae, which keep the leaves on the water surface at all times (1 μm = 0.001 mm)

A third, so-called "waterleaf" is finely divided like a pinna and hardly identifiable as such, because it serves as a "root".

Its extremely high reproductive potential in the form of runners is striking. The daughter plants then form larger coherent chains on the water. If there is a good supply of nutrients, large overgrowths occur in a short time in water arms with little current. Then a closed brownish-green cover forms of this floating fern, which is associated with another small floating fern, the Carolina azolla. The *Salvinia fern* has a morphological feature on its leaf surface that has a similar effect to the lotus effect of the lotus flower (Staeck 2021b, p. 54).

Four tiny, wart-like projections (called *papillae*) sit on the epidermis of all leaves. Their tips are connected in such a way that they look like an eggbeater (Fig. 6.40). This device is used to propel the leaf back to the surface as quickly as possible when a wave pushes it underwater. Thus it can photosynthesize without interruption in the sunlight. If you look at a leaf with a magnifying glass, you can see countless of these papillae as tiny nubs. What you don't usually see, however, are spherical spore capsules sometimes hanging from stems below the water surface.

Among aquarium hobbyists, the eared moss is a popular aquarium plant. Both the family name and the genus name *Salvinia* honors the Italian professor Antonio Salvini (1653–1729). The species name *auriculata* means "ear-shaped." The species *Salvinia biloba* also occurs in Amazonia; however, it is hardly distinguishable from the one described here.

6.4.10 Carolina Azolla (*Azolla caroliniana*): (*Salviniaceae*)

Other Names: Water fern, water velvet, synon. *Azolla cristata*

This mini fern is also ubiquitous in the *várzeas*. The single plant is only 5–10 mm tall and floats on the water surface in stagnant or slow-flowing waters. It spreads rapidly via runners and covers large areas of water—similar to *Salvinia*. On its branched, horizontally running leaf stalks are tiny leaves with papillary hairs on their upper surface to repel water, which overlap in a scaly manner (Fig. 6.41). As the leaves age, they deposit *anthocyanins* (as protection from UV rays), giving them a dark red appearance. Each leaf is divided into two lobes. The upper lobe rests on the water surface and engages in photosynthesis. The lower lobe is thin, nearly colorless, and submerged. Their roots protrude only a few millimeters downward.

Fig. 6.41 The *Carolina azolla fern* is only 5–10 mm in size

They harbour blue-green algae (*cyanobacteria*) in a cavity of the upper leaf that is open at the bottom, with which they live in symbiosis. These fix the nitrogen dissolved in the water, which they supply to the fern plants. This enables the ferns to multiply at such a tremendous exponential rate. Rice farmers in Asia take advantage of this property of the mini ferns. They allow these water *ferns to* multiply en masse among the rice plants, then compost them and use the nitrogen-rich compost material as fertilizer for the rice plants. In Amazonia, this fern is used as cattle and fish feed, and the *Caboclos* use an infusion of the fern leaves to treat sore throats.

The genus name is composed of *azo* (= drought) and *ollyo* (= deceasing) and refers to the dying of the ferns when the dry season sets in. The species name indicates that this plant species occurs naturally as far north as the state of Carolina in the USA. This fern is very popular as an ornamental plant in garden ponds and aquariums. It now has a pantropical distribution and is considered an invasive plant in many tropical countries.

6.4.11 Water *clover* (*Marsilea crotophora*): (Marsileaceae)

This pretty fern plant grows in standing, shallow water where it forms long, slender, branched rhizomes, or it also creeps in the mud when the water level is low. Its alternate four leaflets resemble clovers. The globose spore receptacles arise from the rhizome or the leaf base of a petiole. The family and genus name honors the Italian naturalist Count Aloysius Marsigli (1658–1730). The species name is composed of *croto* (= dirty) and *phora* (= bearing) and refers to the way of life of this plant, which crawls with its rhizomes in the mud.

6.5 Epiphytes

On the margins of rivers and lakes as well as in swamps with up to 12 h of sunlight per day, *epiphytes* (from *epi* = on top and *phyta* = plant) like to colonize the trunks and branches of their carrier plant without harming it. They are in fact squatters: the carrier tree cannot usually prevent colonization unless it lives in mutual dependence with ant colonies that eliminate any incipient growth, as is the case, for example, with *cecropia* (ant trees) (Sect. 6.1.1). Some epiphytes prefer carrier plants in the white-water area and others that grow with a preference on trees of the black-water Igapós. Generally, their seeds are either blown by the wind, as in the case of orchids or transferred to the carrier tree by bird droppings, as in the case of most other epyphyte species.

6.5.1 Ruby-Lipped Cattleya (*Cattleya labiata*): Orchid Family (Orchidaceae)

Orchids are the most diverse family of flowering plants, with over 25,000 species. Around 90% of them are native to the tropics, but some species grow in temperate and even subarctic climates. There are certainly several hundred different species growing in the Amazon lowlands. But when we travel by boat along the riverbanks or hike through the jungle on foot, we see them only by chance, since they usually grow high up in the supporting trees, they are not in bloom at the time, or their often small flowers are overlooked by us. The ruby-lipped Cattleya described here, with its large bright purple flowers, blooms between March and the end of April. The plants like to root with their strong rhizomes (creeping rootstalks) on the branches of weakened trees, which stand in the black water Igapós for several months during the rainy season. Often additional thin aerial roots are formed, which are able to absorb the rainwater running down them. The shoots are swollen (*pseudobulbs*) and serve as a storage organs and water reservoirs. A few narrow lanceolate leaves are at the top of the plant. These are always thick-fleshed (succulent) and parallel-veined as well as smooth-edged.

Fig. 6.42 The ruby-lipped Cattleya is the national flower of Brazil

The flowers are bizarrely shaped. They consist of three outer and three inner petals. The middle of the inner petals is transformed into a lip (*labellum*), which serves to attract the pollinators (Fig. 6.42). Opposite this labellum is usually a club-shaped column consisting of the male stamens and the female pistil, which are fused together. At the upper end of this column is usually a stamen with two anthers in which the pollen is glued together in packets. These are transferred to the pollinator with a "superglue" or by explosion: insects, birds or bats are the pollinators. Below the column is the stigma of the pistil as a shiny replace: Tdisc. The pollen packets brought by the pollinators stick to this. Below the stigma, the ovary is attached below the stigma. The fruits are almost always small capsules containing extremely small and ultra-light seeds: 10,000–20,000 seeds weigh only a single gram. Orchid seeds germinate only when a species-specific fungus infects the seed at a predetermined site ("wicket cell"). It absorbs the nutrients of the seed, the seed now initiates the fungus to die, the nutrients released in this process are sufficient for the seed to germinate.

The name orchid comes from the Greek term *orchis*, which means "testicle". The round root tubers of the orchid moved the Greek natural scientist Theocrat 2400 years ago to this naming. A pair of these herbal "testicles" was considered an aphrodisiac. The genus *Cattleya* is considered the queen of orchids. Its namesake is the English horticulturalist Sir William Cattley, who in 1824 was the first to succeed in getting them to bloom—thus triggering the first orchid frenzy. The *Cattleya labiata* (= lipped) is the national flower of Brazil.

The genus *Epidendrum* (from Greek *epi* (= on) and *dendron* (= wood), i.e. growing "on the tree") is also widespread in Amazonia. Its flowers are often in (false) spikes. By the way, vanilla (from Spanish *vainilla* = small pod) also belongs to the orchid family. The spice vanilla, *Vanilla planifolia, is* originally found from Mexico to Central America.

6.5.2 Parrot Flower (*Psittacanthus cucullaris*): (Loranthaceae)

This plant is a semi-parasite with a woody base and leathery oval leaves that are not dissimilar to our native mistletoe. They take water from the host plant via root-like sinkers, but can also feed independently via their chlorophyll. The sepals and petals are of the same design and between March and May, they form the perianth, which is bright orange-red at the base and orange-yellow at the tip and is about 4 cm long (Fig. 6.43). Hummingbirds, in

Fig. 6.43 The parrot flower is pollinated by hummingbirds

particular, do the pollinating. The spherical, light-green, mucilaginous fruits are picked up by birds. The seeds are skimmed on branches when the beaks are cleaned, or they pass through the intestine and, after excretion, reach tree branches for germination.

It grows on sparse banks of white-water lakes and rivers and in the *várzeas*, where it parasitizes on various woody species. The genus name comes from the term *psittacus* (= parrot) and refers to the lively play of colours of the flowers. The species name *cucullaris* means "hooded" and refers to the bracts which "cover" the inflorescence and the fruit. The family name is composed of *lóron* (= strap) and *anthos* (= flower) because of the strap-like flower segments of other representatives of this family. The plant species described here catches the visitor's eye from a boat even from a greater distance, as its lively shoots and the bright red parts of its flowers stand out well against the green foliage of its carrier plants.

6.5.3 Bromeliads: Pineapple Family (Bromeliaceae)

The incredibly numerous, now more than 3400 described representatives of this plant family are found only in the Neotropics, i.e. from Florida to the virgin forests of the temperate climate zone in Chile. Most of them live as *epiphytes* in the treetops, with their leaves often arranged in funnel-shaped rosettes. In the Amazon lowlands, the three genera *Tillandsia*, *Aechmea*, and *Guzmania* are the most native species. Some of these genera are described hereafter. The family name refers to the Swedish botanist Olaus Bromel (1639–1759).

Tillandsia (*Tillandsia paraensis*)

The genus *Tillandsia* consists of more than 500 species, all of which occur from temperate to tropical South and Central America. In the Amazon lowlands, there are several dozen, which are not always easy to identify, especially since they are usually perched high up on their support trees. They are perennial funnel plants with a compressed axis. The coarse, succulent, parallel-veined leaves are close together in a rosette, with the leaf bases overlapping. This allows rainwater to collect in the funnel. Sometimes the stem leaves also form a false bulb at their base, as in *Tillandsia paraensis* (Fig. 6.44a). In *Tillandsia limbata*, a bizarrely branched red inflorescence rises above the

Fig. 6.44 (**a**) This tillandsia (*Tillandsia paraensis*) likes to grow as an *epiphyte* on dead trees on the banks of blackwater rivers, (**b**) This *tillandsia* (*Tillandsia limbata*) forms a bizarre branched inflorescence, (**c**) This lance rosette lives *epiphytically* and also terrestrially—i.e. "rooting in the ground", (**d**) This Guzmania (*Guzmania sanguinea*) lives *epiphytically* on its carrier tree

grey-green leaf rosette, with bracts resembling thin twigs. The small greenish-white flower tubes develop from their slightly swollen tips. In Fig. 6.44b the flowers are still closed.

The bracts of the inflorescence are often brightly coloured, mostly red or orange and remain attractive for weeks. The predominantly tiny flowers are white or purple. Tillandsias usually live *epiphytically*, rarely terrestrially. They often reproduce vegetatively by means of runners (stolons), whereby daughter plants sprout from the stem of the mother plant. Over years, many shoots of a plant grow close together on the host tree in this way. These rootless plants feed exclusively by photosynthesis, so they are not parasites! The scientific genus name *Tillandsia is in* memory of the Swedish botanist Elias Tillsands (1640–1693).

Lance Bromeliad (*Aechmea distichantha*)

The representatives of this plant genus also mostly live *epiphytically* on trees, gladly sitting on weakening or dead trees. Less frequently they grow terrestrially, i.e. rooted in the ground, like *Aechmea distichantha* (Fig. 6.44c). Because of the compressed shoot axis, the parallel-veined, broad, thick-fleshed, imperforate or even spine-armed stem leaves stand together in a rosette. Thus they form a funnel that collects rainwater. The usually remarkable inflorescence, often up to 1 m tall, is surrounded by crimson coloured bracts, which are strikingly folded back in the species described here. The tiny individual flowers are blue, blue-red, but also yellow. Since they secrete nectar lavishly, they are readily visited by hummingbirds. Representatives of this genus are popular houseplants because they bloom for weeks. The genus name is borrowed from the Greek *aechme* (= "lance tip"), as the foliage leaves are very pointed. The wild form of the pineapple *Ananas nanus* (*nanus* means pineapple in the *Guarani language*), native to the Amazon lowlands, also belongs to this plant family. It grows terrestrially.

Tufted Guzmania (*Guzmania sanguinea*)

The representatives of this plant genus also live *epiphytically* on trees. This genus is only found in tropical Central and South America with more than 200 species. Due to the compressed shoot axis, the parallel-veined, broad, thick-fleshed leaves, which are usually armed with thorns, stand together in a rosette. This forms a funnel that collects rainwater. The tiny flowers are surrounded by coloured foliage leaves. These are lemon yellow at the base and then crimson in colour (Fig. 6.44d). Representatives of this genus are popular houseplants as they flower for weeks.

The genus name honours the nineteenth-century Spanish naturalist Anastasio Guzmán. The species name *sanguinea is* based on the Latin word *sanguis* (= blood) and refers to the crimson coloured bracts.

A related genus is *Neoregelia*, which also occurs in the Amazon lowlands. In this genus, the tiny flowers sit in multiples in the axils of the crimson bracts and partially red-tinted foliage leaves. In other species, the blue tiny flowers sit tightly packed in a head-like inflorescence almost submerged (in the rainy season) in a funnel of water as in *Neoregelia carolinae*, native to the Amazon rainforest. This is also called the "Blushing Neoregelia" because the foliage leaves do not turn red until flowering begins. The genus name honours the

German botanist Eduard von Regel (1825–1892). However, since the plant name *regelia* was already taken, a "*Neo-*" was placed in front of it.

6.5.4 Large Philodendron (*Philodendron maximum*): Arum Family (Araceae)

This mighty plant is evergreen and perennial. Its rhizome-like base becomes woody with age, and the circular light stigmas of the former leaf stalks are clearly visible. In its juvenile stage, it can also climb up the trunks of trees with the help of adventitious roots. In this stage of development, these then entwine around their host tree—like twine (Fig. 6.45). As the years go by, contact with the soil is either maintained, in which case the plant lives *hemiphytically* ("below and above") or it is abandoned and the plant lives *epiphytically*, developing a large number of aerial roots, which, however, hardly reach the soil but take up moisture and nutrients from the air. In older plants, for example,

Fig. 6.45 The large philodendron lives initially as an *epiphyte* developing numerous aerial roots. The cream-coloured flower bulb is surrounded by a decorative bract. The snakelike curved fruit of the philodendron is densely packed with reddish-purple, fragrant berries that attract many species of birds

Philodendron giganteum (with wavy leaf margins), the woody and coarsely spined aerial roots reach the ground, providing additional nutrients. The very large and lobed leaves, hanging from long petioles, are more or less heart-shaped or arrow-shaped, undivided and parallel-veined in both species, with entire margins in *Philodendron maximum*. As the shoot axis of the plant is compressed, the leaves form a rosette of closely spaced and partly overlapping leaves, whereby considerable quantities of withered foliage, twigs, dead insects, and excreta of other creatures accumulate in the course of years. The plant derives many nutrients from this decomposing organic matter. So it can reach many metres in girth over the years. From the ground, this plant looks like a large bush sitting on a branch. Similar to the arum lily (Sect. 6.3.2), the large, first greenish, then cream-coloured flowering bulb is densely packed with innumerable small, whitish, inconspicuous two- to three-petalled flowers, which are sterile below, male above and female at the top. This flower bulb is surrounded by an initially almost closed, later opening decorative, depending on the species creamy white or pale yellow bracts (*spatha*) (Fig. 6.45). This attracts insects in the shady canopies with its bright colour. The flower bulb, which is ready for pollination, raises its metabolic temperature by up to 15°C above the ambient temperature (as in a number of other members of the Arum family, Sect. 6.3.1), which increases the release of scents that attract beetles for pollination. After pollination, this flower bulb develops in some species into a reddish, thin, serpentine fruit on which numerous small reddish-purple, highly fragrant berries are densely packed (Fig. 6.45).

For the family name, see Sect. 6.3.1. With over 4000 species, this family is one of the largest of all, with distribution mainly in the tropics. The genus name *Philodendron* translates as "tree friend", indicating that this genus depends on other trees for survival.

Many *Philodendron* species look very similar and are therefore difficult to identify. The identification is therefore mainly based on the shape of the leaves. Since the leaves are often large, the different species also have corresponding names such as *gigas*, *giganteum* or *maximum*. Besides the species described here, there is a similar species in the same region called *Philodendron undulatum*. The species name means "wavy" and refers to the wavy edge of the leaves. The leaves of philodendrons are readily eaten by red howler monkeys despite their oxalic acid content. The aerial roots are used to make ropes and baskets.

6.6 Vines

6.6.1 Heartleaf Amazon Vine (*Stigmaphyllon diversifolium*): (Malpighiaceae)

Other Names: Butterfly vine, Brazilian glory vine

This evergreen, up to 8 m high fast-growing climbing plant is found by the visitor mainly on the banks of smaller tributaries of the Amazon, but also in the shady forest. The leaves are oval and have a distinct drop tip. The ascending shoots remain rather thin over the years with a diameter of only a few centimetres. They form loose clusters around the branches of their host tree. Loose clusters of bright yellow, orchid-like flowers develop from the leaf axils in April and are very decorative (Fig. 6.46). The five roundish petals, slightly wavy at their edges, merge at their base into a stalk that sits on the flower base. Downward, the flower is terminated by five dark yellow sepals. The ten yellow stamens are barely visible, as they are almost hidden by three leafy, almost vesicular stigmas. The single flower measures only about 3 cm in diameter.

The namesake for this plant family is the Italian professor Marcello Malpighi (1628–1693). The genus name *Stigmaphyllon* refers to the peculiar shape of the stigmas (*stigma* = receptive tip of the carpel and *phyllon* = leaf), the species name *diversifolium* means different-leaved.

Fig. 6.46 The heartleaf vine is a climber with decorative yellow, orchid-like flowers

6.6.2 Red Passionflower (*Passiflora cocchinea*): Passionflower Family (Passifloraceae)

Under good environmental conditions, this climber quickly climbs up its host plant. It has strong, square stems with hard, often glossy leaves that are usually five to seven-lobed; other species also have other leaf shapes. The flowers are solitary on this year's shoot. The basic structure of all passionflowers is always the same.

The base of the flower is generally formed into a nectar container, which is why passion flowers are very popular with birds and butterflies. The most striking feature of the species described here is a wreath of bright red-coloured, fleshy filamentous appendages within the petals, the so-called secondary corollas. This flower component symbolizes Jesus' crown of thorns in Christian mythology (Fig. 6.47a). Equally attractive are the five red petals and likewise five red sepals, which are the same color as the petals, creating a unit of ten petals of the same color. In the centre of the flower, the five yellow stamens sit on a column. The red-stemmed pistil divides upward into three pistils with bulging yellowish stigmas (the stamens symbolize the five wounds of Jesus, the three stigmas the nails with which Jesus was struck on the cross, the pistil symbolizes the wine cup). The ten petals may represent 10 of the 12 apostles—Judas and Peter were absent at the crucifixion. The three (or five) lobed leaves symbolize the lance. The white flower color in other species stands as a symbol of the Savior's innocence, the blue flower color for Our Lady's dress. In addition, each flower fades after only one day. In the axils of the leaves, tendrils sometimes spring up, symbolising the scourges or whips with which Jesus was beaten.

The fruits are also very diverse in shape, but mostly round and greenish-yellow in colour, as in the species presented here. The grey seeds are surrounded by a glassy-white, jelly-like sour seed coat.

The red passionflower is found on the sunny banks of white waters, gladly in floodplain forests, where it climbs up its host plant. It blooms between March and April. Another red-flowered species of the Amazon lowlands, not so frequently found, is *Passiflora vitifolia* (= vine-leaved). It differs from the species described before mainly in having a secondary corolla consisting of two circles of white or dark purple coloured filamentous appendages within the red petals (Fig. 6.47b).

The scientific genus name is derived from *passio* (= suffering) and *flos* (= flower), i.e. the "suffering flower", and refers to the peculiar morphology of the flower (see above). The species name *cocchinea* translates as "scarlet".

This plant was first discovered by Father Simone Parlasca, a Jesuit monk, at Easter 1605 in South America. He presented this filigree flower, rich in

Fig. 6.47 (**a**) The red passion flower likes to grow on sunny riverbanks (*Passiflora coccinea*), (**b**) The passionflower *Passiflora vitifolia* has vine-like stem leaves and a secondary crown of white-purple appendages, (**c**) Passionflower *Passiflora* laurifolia has laurel-like stem leaves and a purple-white banded subcrown, (**d**) Passionflower *Passiflora nitida* has a long, deep blue subcrown with white curled filaments in the last third

symbolism, to Pope Pius V on his return. Later, missionaries used the intricate flower structure to explain the stations of Christ's Passion to the Indians.

I encountered during my boat trips along small white-water rivers two more very decorative species of *passiflora*: *Passiflora laurifolia* (= laurel-leaved, Fig. 6.47c) and *Passiflora nitida* (= glossy, Fig. 6.47d). In general, the visitor has to scan the riparian vegetation very attentively to discover the coloured spots of the *passifloras in* the lush green leaves. Both are also kept in gardens to harvest tasty fruits.

The pulp of passionflowers is often of economic interest in Amazonia, for example, the sweet and sour *maracujá* fruits (*Tupi language* for "food") of the *Passiflora edulis* and the royal grenadilla (*Passiflora quadrangularis*) are offered on the markets.

Passionflowers contain *alkaloids* in their leaves and blossoms, which are often administered in combination with valerian, lemon balm, hops or whitethorn (*crataegus*) so that they have a blood pressure lowering and antispasmodic effect as well as a calming and antidepressant effect and are often prescribed for sleep disorders and restlessness. For example, the drug *Herba passiflorae* is made from the leaves of *Passiflora incarnata* (meaning "incarnate"). In the northern hemisphere, *Passifloras* can be kept and propagated by cuttings without any problems.

6.6.3 Jungle Cucumber (*Gurania eriantha*): Gourd Family (Curcubitaceae)

This climber, which grows to a height of about 8 m, is found on steeper, strongly sunlit banks of white-water rivers. Up to ten decorative, orange-red flowers develop in loose clusters at the shoot ends in April/May. These are actually hermaphroditic, but they develop into unisexual flowers depending on the genetic dominance of the growing mother plant, by either the stamens or the ovules becoming vestigial (Fig. 6.48). All flowers are heavily silky-haired, and the heart-shaped stem leaves are also initially lightly hairy. The five-petaled petals are fused together with the sepals basally cup-shaped. The brown-orange, dermis-skinned fruits, about 6 cm in size, store water and contain bitter substances.

Fig. 6.48 The male flower of the jungle cucumber is densely covered with silky hairs

7

Flowering Plants in the Rainforest

7.1 Introduction

The interior of the Amazon rainforest—the *terra firme*—is a completely different habitat from the sun-drenched banks of rivers or the water bodies themselves. When you enter the primary forest from the riverbank, the light intensity decreases rapidly. After only about 10 m, there is only twilight—comparable to the light conditions about 40 min before sunset in northern latitudes. Usually, only 1–2% of the sunlight reaches the forest floor. Down here, mainly flowerless plants such as mosses, fungi, ferns and lycopods (clubmosses; *Selaginella*) grow, as well as dwarf palms and young plants of various tree species—the "Oskars" (see Sect. 5.3)—and a few shrubs. The blossoms of the trees within the rainforest, however, are mostly in the upper floors of the forest where enough sun reaches. Up there, the trees bloom as well as lianas and *epiphytes*. Flowering trees of the *terra firme* can be discovered especially in clearings. A number of important tree species are presented in this chapter.

The forest itself is not impenetrable, but it is best to have a person ahead of you who will use a machete to clear the planned path. The main purpose of this is to cut off branches hanging in the direction of travel and emerging saplings in order to drive away any dangerous animals, for example, bees, wasps, ants, scorpions and snakes, and to remove unpleasant thorns and spines and trip hazards. Sturdy shoes, long trousers and a shirt with long sleeves are recommended. To avoid getting lost, which can happen very easily, either stay within sight of the riverbank at all times, carry a GPS device or take a knowledgeable local guide with you.

L. Staeck, *Fascination Amazon River*, https://doi.org/10.1007/978-3-662-64452-2_7

Discovering flowers in the *terra firme* is not easy! On the one hand, only a few flowering plants occur on the forest floor—and these bloom normally at the beginning of the rainy season. However, I have more often discovered the decorative red-yellow flowers of the sore-mouth-bush *(Psychotria poeppigiana)* on the dark jungle floor (Sect. 5.3 and Fig. 5.2). Secondly, we are moving directly on the forest floor with a viewing height of perhaps 2–3 m, so our view does not extend far. In a sense, we are directly in the digestive tract of the rainforest, roughly between the "small and large intestine", where all organic material such as plant remains, leaf and flower petals, fruits with their shells, excrement as well as dead animals is decomposed with the help of countless bacteria, fungi and other soil organisms. As a hiker, you will therefore rarely see flowering trees, at best only petals that have fallen to the forest floor. The best opportunity to see flowering trees of the *terra firme* is at the edge of anthropogenic clearings or clearings caused by a forest fire or by a fallen jungle giant, as well as behind the banana, cassava and sweet potato plantations along the shore.

7.2 Deciduous Trees of the Terra Firme

7.2.1 Red Trumpet Tree *(Tabebuia impetiginosus)*: (Bignoniaceae)

Other Names: Ipé, (rosa, yellow trumpet tree; Lapacho—synonym: *Handroanthus impetiginosus*

These mighty trees, up to 30 m high, can sometimes be seen from the river, when they are in full bloom a little higher up in the primary forest (In Fig. 7.1 you see the yellow trumpet tree). Their trunks make up to one-third of the total height, the spreading, long branches about two-thirds. The leaves consist of five finger leaflets attached to the branch by a long stalk. During the dry season, the leaves are shed. At the end of this period, around September, the imposing flowers unfold terminally on the twigs, sitting in clusters, up to 5 cm long and slightly wrinkled in appearance. The single flower is bell-shaped or funnel-shaped and turns into a tube at the bottom. The flower tube contains two longer and two shorter stamens (Fig. 7.2). Related species are the yellow-flowered Ipé (*Tabebuia aurea*; Fig. 7.1), the pink-flowered *(Tabebuia rosea)* and white-flowered Ipé *(Tabebuia alba)*. In the latter, many tubes are reddish veined in the throat. The elongated fruit capsules burst open at maturity with two valves and release seeds, which are surrounded by a papery wing margin.

Fig. 7.1 The yellow trumpet tree flowers in the dry season when the deciduous leaves are shed

Fig. 7.2 The funnel-shaped flowers of the pink trumpet tree

The scientific genus name goes back to the term *Tacyba bebuya*, which comes from the *Tupi language* and means "ant wood". The term *Ipé* is also a *Tupi word* and means "tree bark". The yellowish, hard and heavy, fungus-resistant and thus durable wood of the *Tabebuia* species is sought after and is sold regionally as brazilwood. The species name *impetiginosus* translates as scaly, weeping *(impetigo)* and *contagious (contaginosus)* and refers to a skin rash that is said to be treatable with the plant bark of this genus. To do this, the dried and crushed bark is boiled in water to release the *phenol lapachol* (a vitamin K derivative). The drink is known as *Lapacho tea* and is also used for fever and stomach complaints. However, this infusion is now classified as toxic in our country.

7.2.2 Cannonball Tree *(Couroupita guianensis)*: (Lecythidaceae)

Other Names: Arbre à Bombe ("Bomb Tree"), castanha de Macaco ("monkey chestnut")

This species can also grow over 30 m tall. The leaves, up to 25 cm long and 14 cm wide, are elliptical and sit in clusters at the ends of the twigs. Several times a year the leaves are changed in rapid spurts, which is unusual and for which there is still no valid explanation. Particularly striking for this tree are the many short, leafless lateral branches below the actual tree crown, which attach directly to the trunk. On these sit the flowers and later round fruits the size of a child's head (Fig. 7.3). At flowering time the trunk is covered with flowers. The bizarre, almost porcelain-like flowers, up to 10 cm in size, have six deep red to pink petals. Like all the flowers of the *Lecythidaceae*, the flowers of this species have an unusual shape: the growth of the flower axis on only one side results in a white-pink comb-like structure that curves over the inside of the flower-like a tongue or helmet. On its inner side are hundreds of stamens. The flowers are clustered either in large numbers at the ends of the branches or grow directly out of the stem (Fig. 7.4). This *cauliflory* (that means that the tree is blooming and fructifying directly from its main stems or woody trunks) is more common in tropical trees (for example also in the cacao tree). In the windless lower tree zones, bats take over the pollination of the flowers by clinging to the tree trunk and eating the flower nectar. The flowers give off

Fig. 7.3 The fruits of the cannonball tree, which are the size of a child's head, are one of the favourite foods of wild boars

an intense, sweet scent, especially at night. They are very short-lived and lie almost intact in quantities on the forest floor in the morning. The perfectly spherical fruits, which weigh up to eight kilograms and have a coarse, reddish-brown pericarp—at 25 cm in diameter—harbour an unpleasant-smelling fruit pulp when ripe, which spills over the forest floor when overripe fruits fall down. Large numbers of small fruit flies then buzz around the smelly, pulpy mass. Fruit ripening takes up to 9 months. The tree bears flowers and fruit all year round.

Due to its decorative flowers, this tree species is planted in many Central American countries as a street tree or in parks. The scientific genus name *Couroupita* is the name of this plant in the Indian language of Guiana. The species name *guianensis* also indicates Guiana, where this tree species also occurs naturally. It belongs to the same plant family as the Brazil nut and Paradise nut tree. In the species described here, the fruit has only a suggested lid, which, however, does not open. The hard shell of the fruit is broken by wild boars, which eat the flesh. The leaves contain *alkaloids that* are said to be dream-stimulating.

Fig. 7.4 The bizarre flowers of the cannonball tree are cauliflorous, i.e. they grow directly out of the trunk

7.2.3 Monkey Pot *(Lecythis zabucajo)*: (Lecythidaceae)

Other Name: Cream nut, Sapucaia (Brazilian); synonym: *Lecythis Pisoanis*

This species is one of the giants of the Amazon rainforest, reaching a height of more than 40 m. Its trunk diameter can be up to 2 m. The dense crown is striking. For the natives, such giant trees have a cultural value of their own. If such a tree grows in a village, it is never cut down but respected—also because of its beautiful blossoms and extraordinarily tasty seeds. It sheds its leaves after the end of the rainy season. Then, in September/October, the bright purple, terminal, bizarrely shaped flowers—4–5 cm tall—appear at the same time as the new foliage begins to unfurl. A central purple-tinged petal is enclosed by additional six pale-purple petals. From the central petal emerge numerous yellow-purple stamens (Fig. 7.5). Wood bees specialize in this peculiar flower structure and pollinate the flowers. After a 10-month ripening period, the cup-like fruits, which are up to 30 cm in size and weigh up to 2 kg, are ripe. The lid is blown off, releasing up to 30 seeds (Fig. 7.5). Unlike Brazil nuts,

Fig. 7.5 Like all flowers of the *Lecythidaceae*, the flowers of the monkey pot are also peculiarly shaped. The up to 30 cm large fruits of this tree contain numerous tasty seeds

these are surrounded only by a thin seed coat. However, the seeds are additionally covered with a large fleshy, cream-coloured *aril* (seed coat), which provides an attractive food source for bats and thus contributes to the spread of this species. On the markets of the central Amazon—in *Parintins,* for example—both the empty woody nuts are offered as storage containers and the tasty seeds. Their taste is creamy almond-like with a slight hint of aniseed. The oil of the seeds—similar in taste to corn oil—is also highly prized. However, the seeds absorb and store the *selenium* present in the jungle soil. This trace element is essential for the human body but is highly toxic in higher concentrations. However, the margin between necessary intake and cell damage is very small for *selenium.* It, therefore, seems advisable to consume the seeds of paradise nuts only in small quantities, as the consumer can never be sure whether higher concentrations of this element are present in the soil—and thus in the seeds 10 months later.

In the vernacular, the large nuts are considered a monkey trap. The rumour says if the monkeys reach into the nut—filled with sugar—, they cannot pull out their closed fist through the opening of the nut afterwards and are supposed to be caught in this way.

The Greek genus name *Lecythis* translates as "container" and the species name *zabucajo* is derived from the local name *Sapucaia.*

7.2.4 Cacao Trees *(Theobroma and Herrania)*: Mallow Family (Malvaceae)

The cacao tree needs shade and 2000 mm of rainfall a year to grow and thrive. And these conditions prevail in the *terra firme* where the cacao tree naturally grows as understory plant (besides Mexico and Central America). With its short trunk and a maximum height of 8 m, the tree tends to be one of the small trees and is thus optimally adapted to life under the spreading crowns of the large rainforest trees. The small, five-celled, barely 1 cm large, intricately constructed white flowers are located directly on the main stem *(cauliflorous)* and can be seen almost all year round. The flower consists of five fleshy white sepals, five white petals with reddish streaks inside, five white-yellowish stamens, and a conical yellowish pistil (Fig. 7.6). The petals are pollinated by tiny no-see-ums (midges). If these are absent, for example, because of malaria control or if funnel bromeliads with their water reservoir are absent, fruit ripening fails to occur. After about 4 months, the ellipsoid, 15-cm-long cacao

Fig. 7.6 The cauliflorous flowers of the cacao tree are tiny, but the fruits are surprisingly large at 15 cm. They have a leathery-hard dark yellow to reddish colour

pods are ripe, indicated by the yellow to red coloration of their leathery-hard skin (Fig. 7.6). However, only a few fruits reach maturity due to widespread bacterial and fungal infestation in the humid jungle.

The cacao pod contains up to 60 brownish, flattened seeds, usually called "beans", embedded in a white, sweetish, slimy, tasty pulp. The "beans" are arranged in rows, each about 2.5 cm long. Fresh sweet juice is obtained locally from the pulp *(Suco de Cacao),* the pulp itself also tastes pleasantly sweet.

When Alexander von Humboldt tasted the pulp and raw cocoa in 1799, he wrote in his diary: "No other time has nature crowded such an abundance of valuable nutrients into such a small space as in the case of the cocoa bean in particular." He was truly right about that, without knowing at the time about the proper chemical composition of plant secondary metabolites.

The seeds, together with part of the attached pulp, are subjected to a multi-stage fermentation process on the dry village floor or in a fermentation chamber, in which numerous naturally occurring species of bacteria are involved. These are largely responsible for the later so popular aroma of the cocoa or chocolate. The dry seeds are then filled into bags and exported, for example to the United States or Europe, where they are further processed in chocolate factories. The seeds are ground and liquefied to produce a dark brown cocoa mass. This consists of cocoa butter and cocoa solids. Both components are separated from each other and finally processed into chocolate. The *Caboclos* and indigenous groups also produce raw cocoa for their own use.

Some other interesting ingredients of the cocoa bean should be mentioned here:

- The alkaloid *theobromine* (contained in cocoa powder at 1.2%) was already known to the Aztecs as an aphrodisiac;
- The alkaloid *caffeine* (contained in cocoa powder at 0.2%) has a circulatory stimulating effect (unroasted coffee contains up to 2.5% caffeine);
- The ingredients *epicatechin* and *procyanide* produce the *gamma-valerolactone DHPV* in human body, which has a strong antioxidant effect and counteracts skin aging by slowing the breakdown of collagen. Studies from 2017 show that people who had taken 4 g of cocoa powder daily for 24 weeks saw their existing wrinkles decrease.

In addition to the cacao tree, there are about 20 other species of the genus *Theobroma* that occur from Mexico to the Amazon region. Among these, the large-flowered cacao tree *(Theobroma grandiflorum)* is particularly noteworthy here, as it occurs widely from the lower to middle Amazon and is of regional and cultural importance to the inhabitants of this area. This tree species requires similar living conditions as its famous relative, but grows

Fig. 7.7 The fruit of the *cupuaçu tree*, a relative of the cacao tree, contains an aromatic pulp

considerably taller, reaching up to 20 m in height. The large-flowered cacao tree is not cauliflorous, however, but the stalked, significantly larger flowers are located on lateral branches. The up to 30 cm large, garnet-shaped, brown-red fruits exude a strong, sweetish, pungent scent when ripe. The fruits are similar in structure to those of the cacao tree, and they are sold under the name *cupuaçu* in local markets (Fig. 7.7).

Their sweet and sour, aromatic pulp is just as popular with the local population as the creamy, thick soft drinks and ice cream made from the pulp. Americans and Europeans, however, more often have difficulty with this flavor. According to their own observations, about half of the visitors rave about the tropical-exotic taste of *cupuaçu juice*, the other half categorically reject this drink. By the way, *cupuaçu* butter is used in Brazil to clean wounds due to its antibacterial effect.

The generic name *Theobroma* is composed of *théos* (= god) and *broma* (= food) and thus means "food of the gods". The species name *cacao* is the Hispanization of the Aztec word "*cacauatl*" and refers to the fruit itself.

Tip

A nice souvenir to take home is a body lotion or bar of soap made from *cupuaçu* butter, which contains vitamin E and cell-building derivatives, which are breakdown products of *phytosterols*. In both products, the ingredients of the juice make the skin feel very pleasant, and the fruity-aromatic scent is irresistible and almost tempts you to bite into it. These products can be obtained from Brazilian contacts or in good drugstores.

Finally, there is another relative of the (true) cocoa tree, which grows occasionally in the *terra firme* of the Amazon basin. This is the "Wild Cocoa" (nickname: Monkey Cacao) or *Herrania mariae* which grows in the understory: an evergreen, unbranched, small tree that can grow to about 7 m tall. The small, five-petalled, dark red flowers, about 1 cm in size, which sit in clusters directly on the trunk *(cauliflorous)*, were observed in April/May. From the flower base five white fleshy filaments, each more than 5 cm long, hang out of the flower base, which is very conspicuous (Fig. 5.8). The ellipsoid, 10-cm-long pods, growing on the stem, turn yellowish when ripe. The aromatic, fragrant white flesh is edible and the very fatty seeds can be processed into cocoa powder and then to chocolate in a similar way to "real" cocoa. The generic name *Herrania* was dedicated to the Colombian statesman and general Pedro Herrán (1800–1872).

7.2.5 Lipstick Tree *(Bixa orellana)*: (Bixaceae)

Other Names: Orleans shrub, Bixi, Achiote, Urucu (brazilian)

This small tree, up to about 8 m tall, grows in partial shade. The young twigs have reddish-brown hairs, the leaf veins are often reddish. This tree species shows two varieties: One with white flowers and green seed pods, the other with faint pink flowers and reddish-brown capsules. The flowers, about 5 cm in size, are in panicles at the end of the branches, and each has five sepals and petals, between which are a multitude of pinkish-reddish stamens. The fruits, about 4 cm in size, are softly spiny. Opening them with the fingers the fruit reveals up to 50 seeds, each 5 mm long. These are covered by a scarlet, fleshy seed coat *(aril)* (Fig. 7.8).

Flowers and fruits with different degrees of ripeness can often be seen on the shrub at the same time. The trunk bark contains cork pores *(lenticels)*, which are channel-like openings in the bark that provide the necessary gas exchange with the environment.

In the course of globalisation, this decorative little tree has now become naturalised in many tropical regions, including Asia, as an ornamental and useful shrub.

The genus and family name *Bixa* goes back to the old Indian term "biga" (= red). The species name *orellana* refers to the famous Spanish conquistador Francisco de Orellana, who sailed from Ecuador across the *Rio Napo* to the Amazon and on to its mouth in the years 1539 to 1541.

Fig. 7.8 The seeds of the lipstick tree contain a scarlet dye that plays an important role in the culture of the Yaguas

The seeds of this plant contain the bright red dyes *bixin* and norbixin. The first is fat-soluble, the second water-soluble, which results in many applications in the food and perfume industry. The tasteless dye (trade name *Annatto*) is used to colour butter, margarine, cheese, chocolate, rice, soups, lipsticks and soaps. It also gives wool and silk an orange colour. In Germany, *Bixin* is even recognized as food colouring (E 160b) and can be found, for example, on the surface of the Danish cheese "*Esrom*". Indians, such as the Yaguas, still use it to dye their skin and hair (Sect. 4.1). This is how the name "redskins" came about. A number of South American countries export the dye. The seed coats are also used as an anthelmintic, the leaves and roots as a medicine against headaches.

7.2.6 Pride of Trinidad and Tobago *(Warszewiczia coccinea)*: Coffee Family (Rubiaceae).

Other Names: Wasi Wixi, wild Poinsettia

This tree, which grows up to 15 m tall, likes to stand in clearings where it can stretch upwards between competing trees. It is exceptional for its attractive inflorescences, which is up to 80 cm long, of bright red *bracts* and inconspicuous yellow five petals, only about 6 mm in size, crowded together on the surface of a long axis as a cluster (Fig. 7.9). One to two flowers per cluster have one enlarged sepal to form a long-stalked, elliptical, bright red leaf *(bracteum)* up to 10 cm long. In the green foliage of the forest, these mighty crimson inflorescences are conspicuous even from a distance. The elongated, elliptical leaves, up to 50 cm long, have a drip tip. Strong lateral veins branch off from the leaf midline; the leaf surface is often wavy between them. The spherical fruits, only half a centimetre in size, are two-part capsules containing many tiny seeds.

The genus name refers to the Polish garden inspector Józef Warszewicz, Knight of Rawicz (1822–1866), who first described this plant. Since this name is not easy to pronounce for German and English tongues, it is also often simply called "Wasi Wixi". The species name *coccinea* translates as "scarlet". This plant is the national flower of Trinidad and Tobago, where it always blooms on the national holiday on 31 August. There it is also called "Chaconia" after the last Spanish governor José Chaón. The large red leaves attract hummingbirds, which drink nectar from the flowers. In folk medicine, the plant is used against bleeding.

Fig. 7.9 The unusual inflorescence of the tree "Pride of Trinidad and Tabago" consists of numerous bright red bracts up to 10 cm long, which stand on an axis up to 80 cm long. Between them sit the inconspicuous flowers

7.2.7 Florida Strangler Fig *(Ficus aurea)*: Mulberry Family (Moraceae)

Other Names: West Indian laurel fig, rubber tree, strangler fig

This tree species grows both in the *terra firme* and on the lake and river banks exposed to seasonal flooding. It is not to be confused with the rubber tree (Hevea brasiliensis), which is also native here. It is fast-growing and evergreen and, at up to 30 m, is a mighty tree with a spreading crown. Its way of life is *hemi-epiphytic*, meaning that its seeds, which get onto tree branches with the droppings of birds, germinate there and quickly grow into small trees on their host tree, living *epiphytically* and feeding on themselves; they are therefore not parasites. When the aerial roots of this plant reach the forest floor, their biomass begins to increase exponentially due to the now large supply of nutrients. Numerous secondary stems, transverse branches, and more aerial roots develop until eventually the aerial roots and numerous lateral stems entwine their host tree and eventually cause it to die. Someday, perhaps not for 100 years, this strangler fig will root as a free-standing tree (Fig. 7.10). Further details are given in Sect. 5.9.

Fig. 7.10 Numerous aerial roots hanging down from the main stem as well as thick secondary stems already anchored in the soil are characteristic of the Florida strangler fig. The black colouring of the stems marks the high water line

The up to 20 cm long, short-stalked, elliptical, dark green and very shiny leaves have a drip tip. In all parts of the plant flows a sticky rubber-like milky juice (latex), which the locals use as a substitute for the real rubber, for example for gluing or as chewing gum. The inconspicuous flowers sit in inflorescences containing male, female and sterile flowers. They are spherical in shape and rarely visible from the ground.

All fig species live in a symbiosis with fig wasps; only the wasps pollinate the intricately constructed flowers and, conversely, the female wasps need the flowers to lay their eggs. The fleshy fruits are up to 2 cm in size and spherical and turn golden yellow when ripe. They are eaten by the locals and are also attractive to birds, bats and monkeys, which are responsible for the spread of this genus. In this context, it is interesting to note that the individual representatives of all fig species throughout the rainforest do not flower and fruit synchronously, as is usually the case, but rather each tree develops its own flowering and fruiting rhythm. This ensures that both the fig wasps and the fruit-eating animals are always supplied—and thus contribute to the spread of the figs.

The fig species presented here is just one example of numerous other fig species within the mulberry family that occur in the Amazon lowlands. The appearance of these fig species with mostly mighty downwards hanging aerial roots or enormous root growths at the base of the trunk is similar, but due to the tiny and intricately constructed flowers, an exact identification of the species is difficult. In general, *hemi-epiphytic* fig trees can be distinguished from ground-rooted fig trees. They can reach a great age of well over 100 years.

The genus name *Ficus* translates as "fig", the species name goes back to the first describer, who called this tree species *aureus* (= golden) with regard to the colour of the fruits. The genus *Ficus* includes up to 1000 species, all of which occur in the tropics and subtropics. By the way, our native mulberry is the eponym for this plant family: *Morula* means "mulberry".

7.2.8 Rubber Tree *(Hevea brasiliensis)*: Spurge Family (Euphorbiaceae)

Other Name: Caouchouc (French)

This giant tree of up to 40 m is a typical representative of the *terra firme*. It is not to be confused with the "rubber trees" (*Ficus*), which belong to the mulberry family (Moraceae). The rubber tree sheds its leaves in the dry season. The coveted milky sap (latex), which contains up to 40% rubber-like caoutchouc, flows in the vessels of its bark. The distinctive bark is thin and strikingly smooth and light grey. Also striking are the large, stalked, three-parted stem leaves, which are elliptical with a drip tip. The tiny, inconspicuous, yellowish and penetrating-smelling flowers sit in a large umbel-like inflorescence. You practically never get to see them, however, as they are high up in the tree canopy. Under the trees, you will always find either the 4 cm large, stalked, brownish, three-part capsule fruits or the shiny, 2 cm large, light and dark brown speckled, egg-shaped seeds (Fig. 7.11).

On a hike, through the *terra firme* I once collected some capsule fruits to photograph them later. Upon returning to the ship, I temporarily placed them on the nightstand in my cabin. During the night I was awakened by two explosive pops. Then, in the light, I saw that two ripe capsules had burst open, with the three seeds from each being flung to all corners of the cabin. This is what happens on the trees or on the forest floor, where the fruit capsules bursting at maturity scatter their seeds over a considerable distance. The local

Fig. 7.11 The rubber tree has brown speckled seeds

people like to raise the pretty seeds on chains and sell them as souvenirs. *Hevea seeds* have a high oil content, so they are made into porridge and patties by the *caboclos*. They are also an important source of food for fish, for example the *tambaqui* (black pacu) uses them to fatten itself up (Sect. 8.2.2).

The indigenous people recognized the rubber-like properties of latex at an early stage. Every 2–3 days, a knife was—and still is—used to make a few short diagonal cuts about 1 cm deep into the bark, so that the milky sap vessels are injured but the conductive pathways as a whole remain largely intact. In this way, the bark is successively scratched open to the bottom. The latex then flows into a container, which must be emptied daily to prevent premature clumping *(coagulation)* of the latex (Fig. 7.12). The latex is then heated

Fig. 7.12 The rubber tree's gummy milky sap flows out of the bark from a diagonally applied cut

over fire and smoke, requiring constant stirring, until the mass finally coagulates to form raw rubber (crude rubber). The raw rubber is either dripped onto spinning wooden skewers until the rubber solidifies, or is formed into 40 kg balls and bales. Toys made of elastic natural rubber, such as bouncy balls, have been handed down from the indigenous peoples. Christopher Columbus reported on rubber balls as toys as early as 1495.

Rubber, as is well known, was responsible for the fabulous, incredible rubber boom between 1839 and 1910, which made *Manaus* and *Belém* the richest regions in Brazil and produced many rubber barons. However, every ton of raw rubber was associated with a human life (malaria, worm diseases, malnutrition, etc.). In 1876, however, the Briton Sir Henry Alexander Wickham smuggled several thousand seeds into England. Of these, however, only just under 2400 germinated in Kew Gardens near London, 23 survived as seedlings and formed the basis for large rubber plantations in Mallaka, today's Malaysia, where, however, enormous quantities of raw rubber were not produced until 1910. However, the fairy-tale boom ended in an existence-destroying catastrophe. From 1920 onwards, the US car manufacturer Henry Ford tried to turn back the clock by setting up the huge Fordlândia rubber plantation on 10,000 km^2 about 20 km south of *Santarém*. But in Amazonia's humid climate, such plantations are not viable due to the mass presence of phytophagous bacteria, fungi and viruses. The tubular fungus *Microclyclus ulei* brought this project to an abrupt end.

Tip

The name rubber goes back to the word *cauchum* from the *Tupi language* and means very aptly "the tree that cries". The genus name *Hevea* was derived from the term *hevé,* which the indigenous people used for the rubber tree.

If the reader comes to *Alter do Chão*, a small town 15 km south of *Santarém*, picturesquely situated on a lake of the clear-water river Rio Tapajós, he will find 120-year-old rubber trees scattered all over the place, dating from the time when the transports with the raw rubber from Fordlândia passed by here. Here, for example, between April and May, you can quietly pick up the pretty marbled, shiny rubber seeds everywhere under the trees, which may later remind you of a beautiful trip in a glass on a display case in the living room. In addition, you can also see just germinating seeds as well as older seedlings and saplings up to a size of 20 cm under the enormous crowns of the old *Hevea trees.*

7.2.9 Golden Shower Tree *(Senna spectabilis)*: Legume Family (Fabacea)

Other Names: American Cassia, popcorn tree; synonym: *Cassia spectabilis or excelsa*

This evergreen tree, barely reaching 10 m in height, grows both in the *terra firme* and on semi-shaded riverbanks. It is remarkable because it confirms the theory of continental drift founded by Alfred Wegener. Representatives of this and the closely related genus *Cassia* originally exist both in tropical Southeast Asia and in the Amazon lowlands. They already existed on the large southern continent of *Gondwana* until it gradually split into the now known continents in the southern hemisphere about 100 million years ago. After that, different representatives of the genera *Cassia* and *Senna* developed in Southeast Asia, Australia, Madagascar, South Africa and also in the Amazon lowlands.

The terminal flower clusters of the golden shower tree are pale yellow and not as golden as those of the many well-known *Cassia* species, which are therefore often referred to as golden showers. The single flower consists of five short-stalked petals with the ovary and lower three brown stamens curved upwards (Fig. 7.13). Each petal is elliptic in shape and has a drip tip. The fruits are dark brown, chambered, woody pods about 30 cm long.

7.3 Palms

7.3.1 Peach Palm *(Bactris gasipaes)*: Palm Family (Arecaceae)

Other Names: Prickly palm, Chontaduro (spanish), Pupunha (brazilian)

This fast-growing palm, which reaches a height of 20 m, grows everywhere in the *terra firme.* Characteristic are the broad, ring-shaped bands on the trunk, provided with 5 cm long, pointed spines, which alternate with spine-less bark sections. The end of the palm is a mop of leaflets that resembles the appearance of the coconut palm.

This palm plays an important role for the residents of the Amazon lowlands: the hard wood is used for house construction and the inner core and growing bud is harvested as heart of palm . The palm does not die in this

Fig. 7.13 The golden shower tree of the Amazon basin confirms Alfred Wegener's theory of continental drift

process (as is usually the case) because, it is self-suckering like the açai palm. That means, from adventitious buds on its roots grow new shoots which will develop to multiple stems (so it is also easy to identify), which take the place of the harvested palm. Furthermore, the boiled roots of this palm are used against diarrhea.

The fruit is a 5 cm large, oval drup, which means, an outer fleshy part surrounds a single shell of hardened endocarp with a black seed inside. These drups sit together in large bundles. When ripe, the fruits turn yellow, orange or red, depending on the variety. They can then be seen in April and May

Fig. 7.14 The fruits of the peach palm provide a popular starchy vegetable

everywhere on the markets of the region, where they are sold as *pupunha* (Fig. 7.14).

Birds, especially parrots, also love these fruits. Humans have to cook the fruits for two hours before they can be eaten as a starch- and vitamin A-rich vegetable (like potatoes). The fruits are also processed into fruit juice or protein-rich palm oil. Finally, locals also repeatedly cut down these palms so that weevils settle in the rotting wood, whose tasty, large larvae feed on the dead wood and are a delicacy for people (Fig. 3.2).

7.3.2 Tucumã Palm *(Astrocaryum* ***aculeatum)***: Palm Family (Arecaceae)

Other Name: ***Astrocaryum tucumã***

This palm species is typical of dry *terra firme* sites with open areas and low vegetation. It is considered a pioneer plant that survives wildfires and sprouts again quickly afterwards. However, its seeds take up to 2 years to germinate and it is not until the eighth year that it begins to produce fruit. It is solitary, grows up to 25 m tall and its grey-green leaf fronds stand upwards, as does its inflorescence. Ring-shaped rows of long black spines on the trunk are striking.

Fig. 7.15 The oily flesh of the tucumã palm, which is only a few millimeters thick, has an aromatic, nutty flavor

Its yellowish fruits with a reddish tinge, which are about 4 cm in size and hang in long panicles, are particularly sought after. These are offered at regional markets in April/May, often with the yellow-orange, hard flesh, only a few millimetres thick, already cut off from the thick, wooden core with a knife (Fig. 7.15).

This oily pulp has a pleasant, very aromatic nutty flavor. However, the pulp is also made into ice cream, juice, "wine" and a reddish, clear, high-quality cooking oil (and massage oil). It contains omega-3 fatty acids and the highest concentration of vitamin A and beta-carotene of any known fruit or vegetable, to be precise 52 mg per 100 g of pulp. By comparison, carrots contain only 6.6 mg per 100 g.

The leaf fibers are used to make baskets and hammocks. The thick kernel also provides an excellent almond-like oil for soap, body lotions and shampoos. Its hard wood is traditionally used to make black rings, a symbol of friendship and peaceful resistance to the omnipotence of the state and of the commitment to the poor.

Tip

There are over 300 edible fruits in the Amazon lowlands. But what exactly are they? A simple rule is: hairy fruits, fruits with milky juice and bitter-tasting fruits should not be eaten! Fruits with skin that covers the flesh are hygienically safe.

7.4 Giant Strelitzia (Strelitziaceae)

The last plant species in this chapter is not a tree, but with a height of well over 10 m, it has tree-like dimensions. During my travels in the Amazon lowlands, I encountered this impressive plant species twice, once in the *terra firme* and once on the banks of a white-water river: the giant strelitzia is truly a giant and is therefore mentioned here.

Because of its pretty orange-red-blue flower shape, the South African *strelitzia* is also called the bird of paradise flower. The relative described here—again evidence of continental drift—is a stemless, perennial, slightly woody perennial ("pseudostem") with a stem diameter of 20 cm. The strong rhizomes continually form new shoots, and over the *year*s an enormous cluster of upright shoots is formed. The mighty, up to 3 m long and almost 1 me wide banana-like leaves sit on 2 m long robust stems. A strong, centrally running main nerve (midrib) is clearly visible.

In April/May, the huge erect inflorescence forms, which resembles the *strelitzia* in its structure. It consists of a stem more than 3–5 m long, on which there are several flowers (Fig. 7.16). Each individual flower consists of an immense, more than 40 cm tall, toothed, greenish-yellow bract, from which

Fig. 7.16 The huge greenish-yellow inflorescence of the giant strelitzia resembles the natal strelitzias from South Africa

the acuminate cream-colored petals and sepals protrude. These enclose the stamens and pistil. The stamens are nearly nine inches long, and the anthers seven inches. Each flower produces nectar only once, just before sunset. It is only capable of pollinating for one night, which is done by bats that lick up the nectar with their brush-like, long tongues. As with all representatives of the *strelitzias*, the heavily woody fruit capsules harbor coloured seed appendages *(arilli)*. They hang bright red from the shiny black seeds and attract numerous bird species.

The giant strelitzia is related to the attractive *Ravenala* (Tree of the Traveller) from Madagascar and the large Natal strelitzias from South Africa. All these strelitzias are distant relatives of the banana family. Joseph Banks, Director General of Kew Gardens in London, named this plant family in 1772 in honour of the German Charlotte, born as Princess of Mecklenburg-Strelitz, wife of George III, King of England.

8

Animal Observations by and in the Water

8.1 Introduction

Despite a large number of vertebrate species in the Amazon lowlands, you will see only a few animals during a short stay. But the species presented in Chaps. 8 and 9 are those that can actually be observed during trips along the riverbanks and during hikes in the jungle with the help of one's guides and binoculars. All these animals from the vertebrate families of fish, amphibians, reptiles, birds and mammals, as well as butterflies or their caterpillar stages, I have encountered on quite normal trips by boat or during excursions through the jungle itself.

For millions of years, undisturbed by ice ages, an incredible number of species of both vertebrates and invertebrates have been able to develop in the Amazon lowlands over a huge area of originally around 6.7 million square kilometres with a wide variety of habitats. The table gives an impression of the immense abundance of recorded animal species in comparison with the numbers in Germany (Table 8.1).

Despite these incredibly high numbers, visitors are disappointed to see only a few animals during their—normally short—stay in the amazon basin. On the one hand, this is due to the fact that in and around the water of the many rivers and lakes and also in the huge virgin forest—unlike in Africa, for example—only a few specimen of a particular vertebrate species, such as the howler monkeys, exist The food supply, e.g. edible fruits and leaves or animal prey, is too small for a large amount of individuals. The nutrient-poor soil simply produces too little biomass to supply larger animal populations with food. In addition, the animals living in the ecosystem are exposed to great competitive

L. Staeck, *Fascination Amazon River*, https://doi.org/10.1007/978-3-662-64452-2_8

Table 8.1 Vertebrate species recorded in the Amazon lowlands and Germany

	Amazon Lowlands	Germany
Mammal species	427	107
Bird species	2000	320
Reptile species	378	12
Amphibian species	428	19
Fish species	over 3000	197
Insect species	ca.750.000	approx. 30,000

pressure and therefore constantly roam in the water or forest. Finally, in and around the water as well as in the dense jungle with constantly varying light and shade conditions or very different shades of green of the foliage, each animal is particularly well camouflaged. But from my own experience I can confirm: Every trip has its "animal highlight":

On one of my trips, for example, there were three of the beautifully coloured king vultures that stayed for hours at the carcass of a freshwater dolphin. Another time, during a night excursion, a very large anaconda swam in the headlights so close to my canoe that I could only capture its huge head with my large focal length, leaving the rest of its body out of focus. A last example: on a riverbank, was a large, colourful toucan so busy with its excited and insistent mating calls that we were able to get within a few feet of it.

So there is definitely the chance that visitors of Amazonia can actually observe some of the animals described here. Large animals such as tapirs, jaguars, manatees, giant otters or even the harpy eagle are, however, very rare, extremely shy and have a huge territory, so that you will see them with a lot of luck and exceptionally once if you live in their habitat for a longer time.

8.2 Fish

The diversity of fish life in the Amazon basin is extraordinary! Almost 30% of the approximately 12,000 freshwater fishes described worldwide live in the Amazon basin. Among them, there are specialists such as prey snappers, fruit or flower eaters, scale eaters, remnant munchers, earth eaters and blood suckers, as well as giant fish of considerably more than three metres in length and dwarf fish that barely grow to 2 cm. Moreover there are the fish species whose ancestors were originally at home in the sea, for example, sardines, herrings, pufferfish, garfish (sea needle), flatfish and rays. Near *Manaus*—in the middle of the continent—even bull sharks and sawfish have been caught, which actually live in the South Atlantic. In the meantime, more than 3000 different fish species have been described in the Amazon basin, and the number is increasing almost monthly. In Germany, on the other hand, about 200 species of fish are found, and in the whole of Europe a maximum of about 520. How does this large number come about? There are four main reasons:

1. It is not the productivity of the Amazon waters themselves, but rather the close production-biological interlocking of rainforest and waters. During the rainy season, substantial portions of terrestrial forest products such as leaves, flowers, fruits and seeds go directly into the food cycle of the fish fauna.
2. The annual floods allow a huge expansion of space for the fish. During the rainy season, the normally 300,000 km^2 of water in the Amazon rivers are joined by another 200,000 km^2 of flooded *Várzea meadows* and another 100,000 km^2 of *Igapós* (flood forests). This results in a total area of about 600,000 km^2 during the rainy season, which is more than the area of the US state of Arizona!
3. For many millions of years, undisturbed mixing of fish species has taken place through mutation, selection and adaptation. This means that even in the smallest, sometimes extreme habitats, new species with the most incredible specializations were able to develop (see before).
4. Finally, the three different types of water—white, black and clear—offer more opportunities for the development of new species.

The huge amount of fish is also of great economic importance for the population. Thus, the market value of fish is greater than that of timber, agriculture (for example, cassava, sweet potato) and hunting. This shows that the deforestation of the floodplain forests, which continues on a large scale, endangers the fish fauna to a high degree.

Like catfish and electric eels, for example, many fish species have developed a number of extraordinary adaptive traits as a result of the extreme conditions that prevail in the draining floodplains at the beginning of the dry season, for example with regard to their respiration or their ability to survive dry periods. Furthermore, it has only been known for a few decades that many fish—especially of white-water rivers—are capable of emitting sounds for territorial defence, during stress and also during their courtship rituals. However, we do not usually hear these sounds over water, as 99.9% of the energy is lost when the sound passes into the air. The Viennese zoologist Schaller was the first to use underwater microphones to investigate the various sounds that fish produce underwater. There are twitters, chirps, sustained hums, growls, creaks and trumpets. The sounds are not produced by the larynx and vocal cords, as is the case with us, but by muscle contractions, air movements via the swim bladder, and gill and fin movements. Fish do not have external ears, but they do have a sound-permeable body and an internal ear, so they can hear well.

With the onset of the rainy season, the reproduction of the fish begins. Their reproductive behaviour is triggered by the expansion of their habitats and the drop in water temperature by up to four degrees Celsius due to the cooler rainwater and the associated increase in oxygen content in the river water. Many fish species—ready to spawn—now migrate upstream or into the lagoons and *Várzea floodplain meadows:* where the current is low and nutrient availability is high. The flooded jungle vegetation (shrubs and trees) and the many floating aquatic plants provide shelter and food for the hatched fish larvae. Thus, during the rainy season, a huge food web of plankton, plant food, fish and birds, vegetarian and *carnivorous* animals quickly develops in this habitat due to the incredibly large food supply. It is particularly striking that here, due to the always available abundant food supply, there are dozens of fish species that feed as prey snappers ("predators"), i.e. *carnivorous.* In freshwater habitats of the Northern hemisphere, on the other hand, there are only a few representatives of this group, for example, pike, trout, catfish or perch.

To get an impression of the diversity, colourfulness and specialization of the many fish species, a visit to the fish markets, especially in *Manaus, Belém, Leticia* and *Iquitos* is necessary. But the traveller can also see fish from the Amazon river system in every small village or on the way when meeting fishermen in their canoes. The *caboclos* are always willing to hold up their caught fish or even hand them to visitors for a photo. In addition, a variety of different fish species can be caught in a short period of time using set nets or cast nets, especially on vegetation areas of the riverbank.

Tip

In order to be able to photograph living fish better, it is recommended to have a narrow photo aquarium (for example 30 cm long, 8 cm wide and 25 cm high) made of acrylic (Glass) 0.5 cm thick. There are some specialized companies for this. Since the aquarium is so narrow, larger fish cannot turn around and can be photographed more easily. As a neutral background can serve, for example, blue or gray cardboard.

In the following sections, mainly representatives of the three most common fish groups of the Amazon region are presented: cichlids (Cichlidae), tetras and catfishes.

8.2.1 Cichlids (*Cichlidae*)

Peacock Bass *(Cichla orinocensis)*: *Cichlid* Family *(Cichlidae)*

Other name: Tucunaré (brazilian)

A large representative of this family, which plays an important role as a food fish for the river riparians, is the peacock bass *(Cichla orinocensis)* from the genus *Cichla* (Fig. 8.1). It reaches a body length of over 90 cm, making it the largest cichlid in the Amazon region. *Caboclos* fish for it with large colourful plastic lures (flashers), commercial fishermen catch it with large nets. It is considered the greediest and most voracious hunter found in white and black water (Fig. 8.1). It is offered for sale in every market in Amazonia.

Its basic colouration is golden yellow with three indistinct black horizontal stripes. Its fins are red. Remarkable are its large mouth with deep mouth cleft and conical teeth. This species shows sexual dimorphism, that is, the sexes look different: The larger males have a large, steep forehead or a forehead hump. At the root of the tail is the large black eyespot, which is considered a signal falsification *(mimicry)*. Due to its similarity in shape and colour to a large eye, it either acts as a deterrent to other prey predators or at least distracts them from its head. Predators are confused: Where is in front and behind in this possible prey?

Fig. 8.1 The peacock bass is a voracious predator

The peacock bass is a substrate spawner: the large clutch consisting of several thousand eggs is laid openly, for example on a submerged branch. The female takes care of the brood, the male defends the territory.

A related species that lives only in blackwater rivers is the speckled peacock bass (*Cichla temensis).* This cichlid also has a black spot on the root of the tail but remains smaller. Its basic colouration is rather green-brownish with several rows of transversely arranged white spots.

Marble Cichlid (*Astronotus rubroocellatus*) Cichlid Family (*Cichlidae*)

Other name: (Red Oskar)

This fish species is a popular food fish, which is not missing on any market. It reaches a body size of 45 cm at 1.5 kg and prefers dormant or stagnant white water side arms or lakes, which are only connected to the large rivers during the rainy season. Its large mouth with strongly pronounced lips is striking. The basic colouration of the body is dark green with irregular spots and stripes of bright orange-red and beige colour. This has earned it the nickname "Red Oscar". On the posterior fin rays of the dorsal fin and at the base of the caudal fin are prominent red or orange-red outlined black eyespots, which give it its name (Fig. 8.2). Their function is explained in the description of the peacock cichlid.

Fig. 8.2 In the marble cichlid, the bright orange-red spots and stripes on the sides of the body and the eye dummies on the rear fin rays are particularly striking

Both jaws are equipped with strong muscles and numerous small teeth, which indicates that these perches feed on other fish. The up to 2000 eggs are laid on some substrate. The hatched larvae can cling to the substrate for a few days with a sticky head secretion, even in stronger currents, for example on a stem. After about 5 days they swim away. Their juvenile dress is unusual: on a dark brown ground, there is a whitish stripe pattern. The fish are very popular with aquarists who specialize in large cichlids.

Flag Cichlid *(Mesonauta insignis)*: Cichlid Family (*Cichlidae*)

This up to 15 cm large fish species lives in the bank vegetation of small white water rivers with a low flow velocity. The characteristic broad black stripe runs diagonally upwards from the mouth over the eyes to the front fin rays of the dorsal fin. Usually, this perch stands with its body slanting upwards just below the water surface, so that the stripe now appears to run horizontally and its pattern of markings continues in the surroundings and the body outlines become blurred in the moving water—a good example of *somatolysis* (= dissolution of the body). Thus this fish is optimally camouflaged against predators. It is *omnivorous*, i.e. it eats the plant and animal food, such as plants, fish eggs, small fish, insects or small crustaceans.

Flag cichlids are open breeders: they lay the more than 300 eggs directly on leaves, submerged branches and tree trunks in the flood forest, where the clutch and later the hatched young fish are guarded. This fish species is popular with aquarists because both parents guard and guide the young.

Demon Eartheater *(Satanoperca jurupari)*: Cichlid Family (*Cichlidae*)

These cichlids, also popular with aquarists, grow between 14 and 27 cm. They prefer calm bays with sandy substrate and branches in white water streams.

Their basic colour is beige or light gray. In males, all fins and the ventral side are bright yellow to orange. The dorsal fin has numerous fin spines (be careful when touching!). Their last fin rays, the ventral fins and the anal fin are elongated.

Their feeding strategy is to sift through substrate (e.g., sand, small rock, silt) for food, where mouthfuls of substrate are taken up and sorted in the buccal cavity for food items as insect larvae, snails, worms, rotten plants. Therefore they are called *geophages* (= eartheaters). The several hundred eggs are laid openly on the ground, after which both parents or only the female take care of the clutch. The eggs are taken into the mouth where the larvae hatch—they are true mouthbrooders. The young fish are only released into the independence of the open water when their development is complete.

The generic name *Satanoperca* is composed of *satan* (= devil) and *perca* (= perch), hence its name.

Heckel's Cichlid *(Acarichthys heckelii)*: Cichlid Family (Cichlidae)

This cichlid resembles the demon eartheater in size and shape, also its habitat is similar. It is omnivorous. The mating colouration of the male shows olive green sides with bright orange areas extending to the head. The fins then also turn orange (Fig. 8.3).

The several hundred eggs are glued to the ceiling of a cavity. The female cares for the clutch, the male guards the territory. The fry hatch after a few days and are cared for and guarded until they are independent.

The genus name *Acarichthys* is composed of *acará* (= *Tupi language* for cichlid) and ichthys (= fish), the species name honours the Austrian ichthyologist Johann Heckel.

Fig. 8.3 The Heckel's cichlid is a popular aquarium fish

Leopard Pike Cichlid *(Crenicichla lenticulata)*: Cichlid Family (Cichlidae)

This pike-like, elongated, up to 40 cm large fish is characterized by its marbled body side and dotted head. Behind the gills is a faint reddish spot. The dorsal fin extends almost over the entire back. It likes to inhabit fast-flowing blackwater and white-water rivers. We were able to catch it and other fish described here easily with the dragnet (Fig. 8.4a).

Fig. 8.4 (**a**) The leopard pike cichlid is a voracious prey snapper, (**b**) At vegetated bank sections or in the dry season, numerous fish species can be caught in a short time with towed nets

This pike cichlid is a voracious prey snapper which was even startled more often by the sound of our boat's engine and jumped into our zodiac. As a cave-breeder it sticks its clutch for example to the ceiling of a tree cave, the male defends the territory. The young fish are supervised by both parents.

The genus name *Crenicichla* is composed of *krene* (= comb), which refers to the serrated gill cover, and *cichla* (= denotes this genus of fish). This genus comprises more than 80 species, most of which live in the Amazon lowlands.

8.2.2 Characids

The family *Characidae* is within the characiforms very rich in species and includes more than 1100 species worldwide. In the Amazon basin, there are morphologically and physiologically very different species of characiforms including many carnivorous, but also numerous purely vegetarian species that eat fruits, seeds, leaves and flowers during the rainy and flood season. During this time, birds and also monkeys actually have to share the flood forest habitat with fish. Many of these herbivores are schooling fish.

By the way, most of the characidae—like the *Salmonidae* (for example salmon and trout)—have an additional small adipose fin, which is located between the dorsal and caudal fin. Finally, characids have the so-called Weber's apparatus—ossicles that serve to transmit sound waves from the swim bladder to the inner ear. This allows them to hear excellently.

Many of the fish described below are considered popular food fish. Due to the overfishing that has been going on for years and the constantly increasing population in the entire Amazon basin, many of these fish species are now threatened in their existence. On the fish markets visitors will notice boards, which prohibit the catch of the affected species between October and March of each year under penalty, but unfortunately, no one adheres to it and the authorities are generally considered corrupt in Brazil.

Red (-Bellied) Piranha (*Serrasalmus nattereri)*: Characin Family (*Characidae*)

Synonym: Pygocentrus Nattereri; Piranha preta (brazilian): "true piranha"

This fish is certainly the most famous species of the entire Amazon region. Its name is associated with many anecdotes, sensational reports and adventure film about its dangerousness and aggressiveness. Even Alexander von Humboldt wrote in a diary entry from April 3, 1800: "It attacks people while bathing and swimming and often tears off respectable pieces of flesh ..." And U.S. President Franklin D. Roosevelt recorded in his 1914 report, "They tear and devour alive every wounded man and animal; for blood in the water brings them to frenzy ..."

Adult piranhas have razor-sharp teeth which, in combination with extremely strong jaw muscles, can also be dangerous to humans if frightened animals are handled carelessly—for example, if you try to detach them from the fishing hook. Numerous accidents have been documented in which deep flesh wounds occurred or a fingertip was torn off or a toe was mauled by a piranha that was bouncing around on the bottom of the boat after being released from the fishing hook. Nevertheless, both the natives and we as visitors like to swim without problems in waters where there are many piranhas.

While the red piranha grows to only about 20 cm (Fig. 8.5), the black or red-eyed piranha *(Serrasalmus rhombeus)*, which occurs exclusively in black water, reaches up to 40 cm with a weight of up to 2 kg. Thanks to its particularly large jaw muscles, this fish in particular bites with 30 times its weight (Fig. 8.6). Its enormous bite force is even stronger in relation to its body weight than that of a great white shark or caiman.

In the red piranha, a strong orange-red zone extends from the chin over the chest to the abdominal region. Behind the gill covers, it extends to the shoulder area, the rest of the body is silvery-grey. Except for the dorsal and caudal

Fig. 8.5 The most famous fish in the Amazon is the red piranha

Fig. 8.6 The black piranha weighs 2 kg and bites with 30 times its own weight

fins, all fins are of intense red colouration (Fig. 8.5). Juveniles show a conspicuous pattern of black spots.

It is mostly found in white water rivers and lakes in a wide variety of habitats where it is extremely common. Like all characidae, it has a well-developed adipose fin and a small spine just before the beginning of the dorsal fin. The lower jaw is bulgingly protruding. The teeth, which are in a row in both jaws, have a large, sharp triangular point in their center and two much smaller ones on the sides. This enables it to cut pieces of flesh out of its prey as if with a scalpel. When fighting over prey, piranhas emit drum sounds.

In the live fish, the teeth are hidden behind the lips, so that only the tips of the teeth can be seen. Red piranhas eat mainly fish, but also snails and insects, as well as plant food, such as seeds and fruits that have fallen into the water, which is proven by studies of the stomach contents. Piranhas practice an intensive brood care, so they defend the clutch against predators, even against humans.

The name "serra" (= "saw") in its species name does not refer to the powerful dentition, but to the toothed, keel-like and therefore saw-like underside of the body, on which the scales are transformed into sharp-edged spines. So the genus name *Serrasalmus* refers to this morphological peculiarity: *Serra* (= saw) and the second part of the genus name *salmus* means "salmon", because the morphology of both groups is similar. The species name honors the Austrian naturalist Johann Natterer, who drew the red piranha in 1828 during his 18-year Austrian Leopoldina expedition. The name piranha in turn comes from the *Tupi language* and is composed of *pira* (= fish) and *nja* (= the one that bites), thus very aptly: "The fish that bites".

Large piranhas are regularly offered at local markets. More than 25 different piranha species occur in the Amazon basin, including scale- and fin-eaters such as the wimple piranha *(Catoprion mento)*, pure vegetarians that eat only fruit, and very special food specialists—such as a piranha species, which feed mainly on the caterpillars of a butterfly, which in turn eat the poisonous leaves of the small riverside piranha tree (*Piranhea trifoliata,* family Euphorbiacae). The flesh of these piranhas is said to taste so vile that cats immediately spit it out.

Tip

With a simple rod made of a string and a hook with bait (a piece of fish or meat), even inexperienced anglers can catch dozens of piranhas of different genera in a short time in quiet locations near a river bank. But be careful when removing the hook! This is better left to experienced local boatmen.

Being stressed, the fish make sounds like the soft "woof, woof" of a dog.

By the way, grilled piranhas are tasty. You can have the dentures prepared and take them home as a souvenir.

Pirapitinga (*Piarachus brachypomus*): Serrasalmid Family (Serrasalmidae)

Other names: False Piranha, red-bellied Pacú

The juveniles of this fish species are easily confused with the red piranha, as they have a similar body shape and are red in colour, especially on the chest and belly (Fig. 8.7). The *Caboclos, therefore,* call it *Pirapitinga,* which means "false fish" from *pira* (= fish) and *pitinga* (= false). This is because when young, these fish imitate the behaviours of piranhas to deter predators through *mimicry.* As an adult fish, it then exhibits a dark, dirty-white-blue colouration.

Fig. 8.7 The pirapitinga is called "false fish" by residents because, as a juvenile, it is strikingly similar in colour and shape to the red piranha

Even if you look closely at this fish, you can tell by the teeth that it can't be the red piranha: They are compact molar teeth with a slight overbite. They feed in the floodplain forests on fruits and seeds, for example, those of the rubber tree, which have fallen into the water.

This fish species becomes very large with a body length of 80 cm and a weight of up to 20 kg and is very popular among the population as a food fish.

Black Pacú *(Colossoma macropomum)*: Serrasalmid Family (Serrasalmidae)

Other names: Black-finned Pacu; Tambaquí (brazilian)

This impressive fish is the largest species in this family with a body length of up to 1.10 m and a weight of up to 40 kg. Its basic body colouration is slate grey, but there are local colour variants in which the areas between the lower jaw and the pectoral fins are orange and the gill region is red (Fig. 8.8). There are also regional subspecies that are even more colourful.

These fish tolerate low oxygen levels, as occurs in drying waters in floodplain forests. They also spawn there between September and February. The young fish feed carnivorously on plankton, insects and crustaceans, while the adult fish live purely vegetarian. They do this by swimming into flooded forests, where they eat submerged leaves and crack open hard-shelled nuts. Their dentition resembles that of cattle, but the teeth are even in two rows one behind the other (see Fig. 8.8). They are particularly fond of the seeds of rubber trees. The sound of the seeds hitting the water surface and the etherical

Fig. 8.8 The black pacú is an impressive fish, with up to 1.10 m in body length and weighing 40 kg. Its teeth sit in the lower and upper jaw in two rows behind each other

oils and *alkaloids that* escape attract them in large quantities: They have an excellent sense of smell with numerous sensory cells around their mouths. During this time, they accumulate eat large fat reserves. Hard palm fruits are swallowed whole, the pulp digested and the hard-shelled seeds excreted as germinable seeds after passing through the intestine.

This fish species is an economically important food fish in the Amazon basin and can be found in all markets. The boneless fillet, which contains up to 10% fat, is extremely tasty but expensive—definitely recommended and can be found on the menu under the name *Tambaqui*. Between October and April, this type of fish may not be caught commercially.

The name Pacú is of Indian origin.

Silver Dollar *(Metynnis argenteus)*: Serrasalmid Family (Serrasalmidae)

This silvery-coloured tetra lives in calm white and clear water rivers. It grows up to 30 cm, is strikingly high-backed and laterally compressed. The scales are particularly small. The dorsal fin is short, but the anal fin is seam-like long, with an orange-coloured fringe. Juveniles have a round, circular black spot on both sides of the body. They are omnivorous and eat plant parts as well as insects. They are offered as food fish in all local markets.

The species name *argenteus* means "silver" therefore also its English name.

Flag-Tailed Prochilodus *(Semaprochilodus taeniurus)*: (Prochilodontidae)

Other names: Striped Tetra, flag-tailed Tetra; Jaraqui (brazilian)

This fish species reaches a body length of about 30 cm, the elongated, bump-like, silvery-coloured body ending in a deeply forked caudal fin with the characteristic alternating yellowish-white and black stripes (Fig. 8.9).

On the lips are two rows of very small comb-like teeth, with which the animals scrape-off algae, microorganisms and *detritus* from solid surfaces, for example from tree trunks in floodplain forests. The latter, solid or semi-solid organic remains that are already undergoing bacterial decomposition, represents an important component of the food spectrum of these fish.

They live in shoals in flowing, smaller white water rivers, where they are crepuscular and nocturnal. With the onset of the rainy season, especially when the water rises sharply in April, the gonads of the fish begin to produce enormous quantities of eggs and sperm. In females, eggs then account for

Fig. 8.9 The flag-tailed prochilodus is an important food fish in the *Manaus* region

up to 17% of their body weight. A female produces up to 300,000 eggs! Now is also the time when these fish gather in huge shoals of thousands and migrate up to 1000 km into the upper reaches of the rivers. There, especially at full moon—as if on command—eggs and sperm are released together freely over water plants, accompanied by a deep humming of the male fish, sounding like a motorcycle at idle speed. At least this is what expedition members reported in their diary notes. This "underwater chorus" can also be heard as a diffuse background noise outside the water. The eggs and the hatching fish larvae reach the fertile floodplains *(várzeas)* with strong currents during the rainy season.

The riparians of the rivers catch these fish in enormous quantities during the time of their ascent into the upper reaches. The large quantities of eggs and sperm in their bodies make them extremely rich in protein. They can be found for sale in all regional markets.

Dogtooth Characins (Cynodontidae)

Other names: "Cachorro" (portugese for "dog", freshwater Barracuda)

These streamlined, prey snappers are found in fast-flowing white and black waters like creeks and *igarapés,* where they ambush smaller fish. These fish are living fossils, as they have been around since the *Miocene,* millions of years ago.

The silvery **biara** (*Rhaphiodon vulpinus)*; dogtooth characin family (Cynodentidae), for example, is such a prey snapper with slightly bluish-yellow areas over the head and gill areas. It reaches a body length of up to 1 m. Its dorsal fin is located in the rear third of its body. In case of danger, it also jumps out of the water—for example, it jumped into our canoe twice. Its long pectoral fins and above all its two elongated, sabre-toothed fangs in the lower jaw are striking.

Another predator is the **spotted cachorro** (*Acestrorhynchus falcatus)*; (Acestrorhynchidae). This slender fish, also known as the freshwater barracuda, remains smaller than the sabre-toothed tetra at about 40 cm. Its large sickle-shaped dorsal fin is striking. Its basic body colouration is metallic green-reddish. Its large, deeply cleft mouth displays numerous conically shaped fangs (Fig. 8.10).

Both fish species are easily fished with flashers.

Fig. 8.10 The "cachorro" (Brazilian for "dog") has a strong dentition of conically shaped teeth

8.2.3 Catfish

This catfish family is also very rich in species, with over 3000 different species worldwide, most of which live in freshwater. The diversity of forms and lifestyles is therefore unbelievably large and ranges from the barely 3-cm large bloodsucker *Candiru* to the almost three metres large Central European wels catfish (sheatfish). Furthermore, there are also detritus eater, and carnivore species.

Most conspicuous characteristics of the catfish are their prominent barbels, whisker-like sensory organs near the mouth, which serve to touch and taste their food. Their skin is often scaleless, but armored with bony plates. Many have spines, which are located directly in front of the pectoral fin or the dorsal fin and are intended to deter predators, for example cormorants .

In addition, many catfish are also vocal: they can emit acoustic signals, for which they have developed various methods—there are, for example, gill cover or pectoral muscle sounds as well as vocalization via fin and gut movements.

Sailfin Catfish *(Pterygoplichthys multiradiatus)*: Armoured Catfish Family (Loricariidae)

Other names: Suckermouth Catfish, common Pleco

The armoured catfish family includes more than 700 different species, many dozens of which are found in the Amazon region alone. The sailfin catfish have a flattened, elongated, up to 40 cm long body, which is armoured all over with bony shields. Particularly striking is their sail-shaped, large dorsal fin, which, like the pectoral fins, has razor-sharp fin rays (Fig. 8.11). Nevertheless, cormorants can swallow smaller catfish whole. On several occasions I have observed them throwing the fish up, again and again, turning it several times until it is finally in the right position to swallow. During—this enormous stress these catfish often make grunting noises.

They live in white waters, but they prefer flooded meadows *(várzeas)*, oxbow lakes and calm lakes with a muddy bottom. Their body colouring is black-brown with similar patterns of spots and dots. In addition to short barbels, their mouth has strong lips that are formed into a suction disc. They use them to rasp off algae growth, for example from tree trunks. They also burrow in the muddy substrate, from which they sniff out worms, insect larvae and other bottom-dwelling animals.

For reproduction, they dig a shallow pit in which eggs are laid. The males engage in brood care. When lakes and oxbow lakes dry up in the dry season,

Fig. 8.11 Sailfin catfish are armoured all over with bone shields

they are able to cover greater distances by land to the nearest watering hole with their powerful tail fin and pectoral fins.

They are caught with trawl nets and offered as food fish in large quantities on the local markets. Their meat, however, is considered inferior.

The complicated genus name *Pterygoplichthys* is composed of three parts: *pterygion* (= fin), *hoplon* (= weapon) and ichthys (= fish).

Redtailed Catfish (*Phractocephalus hemioliopterus)*: Long Whiskered Catfish Family (Pimelodidae)

Other name: Pirarara (brazilian)

This imposing catfish has a massive body with black ground colouring. In addition, a whitish or yellowish inconsistent band runs over the entire side of the body. Striking are its orange-red tail and its equally coloured dorsal fin, which gave it its Brazilian name Pira-Ara: *pira* (= fish) and *ara* (= the red macaw parrot). Two pairs of barbels up to 15 cm long serve for orientation (Fig. 8.12). It is gifted with a loud voice and emits trumpet-like calls by forcing water and air with high pressure through the muscle-reinforced gill covers (e.g. when hanging on a fishing rod).

It is nocturnal and eats crabs, small fish, but also fruits that have fallen into the water. It grows up to 1.3 m long and can weigh more than 40 kg.

Fig. 8.12 The redtailed catfish is one of the impressive fishes of white water, with a body length of up to 1.30 m

As a food fish it has only local importance, but from time to time it can be found in the nets of the *Caboclos.* As a colourful juvenile fish it is often kept in the aquarium.

Its generic name *Phractocephalus* is composed of *phractos* (= fence) and *kephale* (= head).

Tiger Surubim (*Pseudoplatystoma tigrinum*): Long Whiskered Catfish Family (Pimelodidae)

Other name: Surubim (brazilian)

This large nocturnal catfish is a lurking hunter that eats fish. It grows up to 1.3 m long and reaches a weight of 30 kg. Its whitish ground colouration has a longitudinal black tiger-like colour pattern on the back and sides that varies among similar species (Fig. 8.13). Its spatulate mouth and particularly long antennae-like barbels are striking. It is one of the few fishes to live in the main Amazon River, but also in other white-water rivers at greater depths and well camouflaged, for example in sunken trees, where it waits for prey. With the onset of the rainy season, it follows its prey range.

Its tasty, almost boneless meat makes it a popular edible fish, which is caught in gillnets and offered every day on the markets. It is now threatened by overfishing despite a closed season for many months.

Fig. 8.13 The tiger-like colour pattern of the tiger surubim is characteristic of this popular, boneless food fish

Ripsaw Catfish (*Oxydoras niger*): Thorny Catfish Family (Doradidae)

Other name: Cuiú-Cuiú (brazilian)

This catfish has a stocky, black body with a wide mouth and rump at a body length well over 1 m. Its bag-shaped mouth is shaped into a sucking mouth that seeks out detritus and hidden insect larvae and crustaceans in the muddy bottom of calm white-water igarapés and lakes. Several pairs of barbels on its lower jaw help it do this. Its dorsal fin and pelvic fins are equipped with scalpel-like, toothed or grooved spines. Its skull is highly ossified and the flank spines arranged in a row on each side of the body are particularly striking (Fig. 8.14). With this equipment, it is well protected from caimans and even jaguars, for any stomach would be lacerated by swallowing these sharp thorns. This carapace is said to withstand even machete blows.

The fish is crepuscular and nocturnal, during the day it lives hidden between plants and roots. When stressed, it emits a muffled grunt with the help of its muscular, sharp-edged pectoral fins. *Caboclos* catch it with nets, so you can photograph it here and there.

The genus name *Oxydoras* is composed of *oxys* (= sharp) and *dora* (= skin).

Fig. 8.14 With its flank spines and scalpel-like ventral fin spines, the ripsaw catfish (Cuiú-Cuiú) is even safe from jaguars and caimans

Candirú *(Vandellia* spec.*)*: Whale Catfish Family (Trichomycteridae)

Other names: Vampire fish, Canero (spanish)

The term *candirú* (*Tupi language* for "little devil") refers to fish species of different genera and families that feed parasitically on the blood of larger fish. They live on sandy bottoms of small white-water rivers with low currents or in the foliage at the bottom of black-water lakes, waiting for scents of urea released by passing fish through their gills or intestinal openings. Among these candirús there are very small, worm-like species, barely reaching 3 cm in length.

They have needle-shaped teeth which—when they smell the urea—spiral with their rotating body into the gill leaflets or the intestine of the host fish. Small hooks behind their own gills improve their retention in the gills or intestines of their host. They perforate the vessels of their host with their teeth to get access to the blood. After the meal is complete, they leave their food donor.

I caught the candirú presented here, only about 3 cm long, between wilted leaves in 20 cm water depth on the shore of a blackwater lake. Others—over 20 cm large—we have fished with bloody pieces of meat. The species presented here has a scaleless, cylindrical body and a remarkably long fin seam on both sides. Six barbels enable it to orientate itself. Its reddish colouration camouflages it perfectly at the bottom of the tea-coloured blackwater among dead leaves (Fig. 8.15). The tiny eyes perched above the mouth are barely visible. This fish likes to swim at smaller fish to enter their intestinal opening. There they attach themselves to the intestinal wall, where they suck blood through the existing blood vessels.

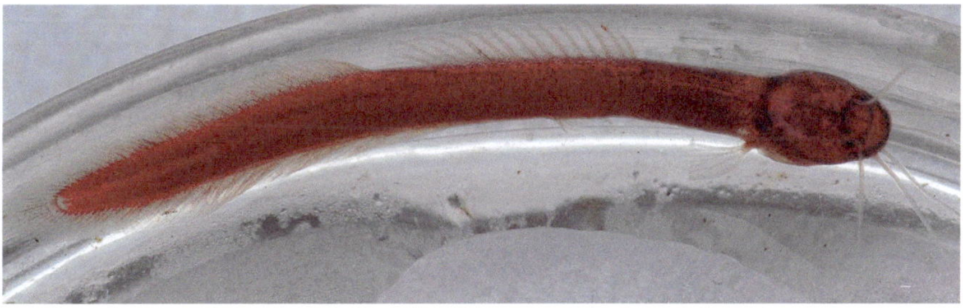

Fig. 8.15 Candirús invade the intestines of their host, where they feed on their blood. It also happens that they enter the body orifices of humans

Reports from *Caboclos* indicate that small candirús are also attracted by the menstrual blood of bathing women as well as by human urea, such as people urinating in the water, which is a traditional practice of river riparians. The candirús are said to have already invaded the vagina, anus, noses and ears. Their removal is problematic because of the barbs on the gills of these fish.

The species of the fish shown in Fig. 8.15 cannot be clearly identified. The family name refers to the whale-like body shape of many representatives of this family.

8.2.4 Silver Arowana *(Osteoglossum bicirrhosum)*: Bony Tongue Family (Osteoglossidae)

Other name: Aruanã (Brazilian), water monkey

This elongated, ribbon-shaped, laterally flattened fish species grows up to 1-m-long and has a silvery ground colour. The animal swims in a peculiar wave-like manner. There are two brownish-green barbels about 4 cm long on the lower jaw at the oblique mouth opening (Fig. 8.16). Their upper jaw is equipped with many small, pointed teeth. The long anal fin extends to the caudal fin. Its eyes are divided for underwater and above-water vision.

The fish prefers oxbow lakes of white-water rivers, where it snatches insects, frogs and lizards sitting on branches above the water, for example, with targeted jumps out of the water. I have observed this catching behaviour several times, once an arowana jumped continuously out of the water to catch a butterfly gliding over the water. In doing so, he can jump up to 2 m high or far, respectively, which has earned him the nickname "water monkey" among the locals. It has also happened several times that he jumped into our canoe from fright.

Fig. 8.16 The silver arowana ("water monkey") jumps up to 2 m out of the water to snatch its prey on branches near the shore

These fish are also mouth brooders, which means that after laying eggs in a sand pit, the male takes the 1.5 cm large eggs into his mouth, which the hatched fry only leave after about 50 days. Then they are already 10 cm tall.

The arowana is a popular, tasty edible fish that is always offered at regional markets, often rolled into bundles. The generic name *Osteoglossum* means "bone-tongued" and the specific name *bicirrhosum* means "two barbels".

8.2.5 Cururu Stingray *(Potamotrygon wallacei)*: (Potamotrygonidae)

In the Amazon basin, there are about 30 species of stingray recorded so far, with the largest reaching a body diameter of 1 m and a weight of 200 kg. Generally, they live in slow-flowing white-water streams, lagoons or in oxbow lakes near sandbanks. There they usually lie motionless and well camouflaged in shallow water, lurking for prey such as insects, snails, crustaceans, small fish. The species presented here is of almost chocolate brown colouration with countless small irregular light brown spots and few larger black edged circle-like patterns. This makes these fish perfectly camouflaged in the loamy yellow water.

Their body is a nearly circular disc about 40 cm in diameter. The gills are located on the ventral side and the paired breathing holes are located behind the small eyes. The strong whip-like tail is toothed like a saw with strong, sharp spines that are barbed. They are provided with longitudinal grooves over which the venom is passed, which is formed in glands at the base of the spines. The teeth are blunt. The dorsal and caudal fins are absent (Fig. 8.17).

Their reproduction is special: the yolk-rich eggs are not laid on a substrate but are nourished in the mother's womb for about 3 months until the development into a young fish is complete. Then the about 8 cm long young fish are born alive and are immediately independent.

The ray shown here was caught with a trawl net during the dry season in water only 30–40 cm deep. While still in the net, the female gave birth to some not yet fully developed juveniles (Fig. 8.17).

The tasty meat of the rays is occasionally offered for sale at regional markets. People in the Amazon lowlands fear stingrays because they are difficult to spot when they are lurking on the ground. If they feel threatened, for example, if you step on them, they can cause serious wounds and poisoning with their tail strokes. If an artery is hit, the injury can even be fatal. Their venom can also cause blood poisoning: The Austrian naturalist Johann

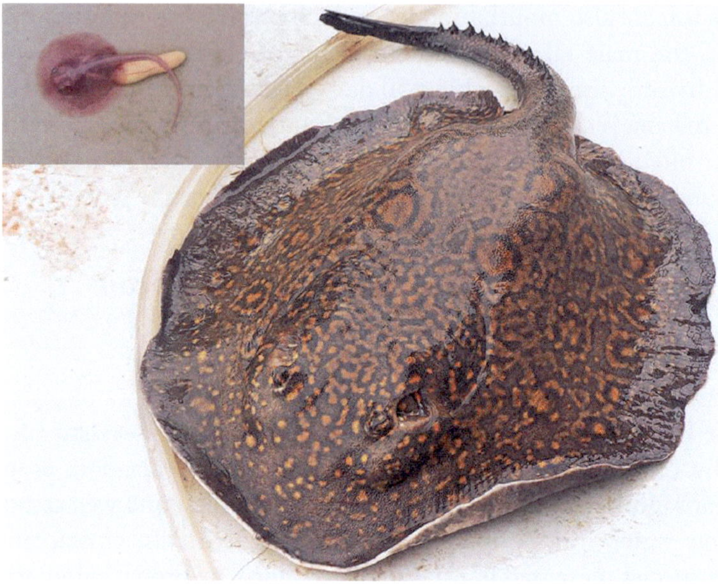

Fig. 8.17 The whip-like tail of this freshwater stingray cururu has barbed spines. In case of danger, they release a life-threatening poison via these spines. The small picture shows an incompletely developed stingray with a large yolk sac

Natterer already drew and explored the freshwater ray on his expedition in 1820. A member of the expedition, Hannes Zell, was severely injured by the poisonous sting of a stingray. Brazilian doctors even wanted to remove his festering leg. The painful healing then took over a year.

8.2.6 Electric eel *(Electrophous electricus)*: Neotropical Knifefish (Gymnotidae)

Despite its elongated, eel-like appearance, the electric eel is not an eel, but a knife fish. Its body is reddish-brown or grey and can reach a body length of over 2 m with a weight of more than 15 kg. The anal fin runs as a narrow fin seam almost over the entire body and ends at the tip of the tail. Dorsal, caudal, and pelvic fins are completely absent, and the small pectoral fins are paddle-like (Fig. 8.18). The flattened head has a wide toothless mouth that can accommodate frogs as well as small fish, insect larvae, and crustaceans. In addition to gill breathing, which provides about 20% of the oxygen it needs, numerous fanned, heavily perfused papillae (wart-like protuberances) sit on the roof of its palate and are capable of taking in atmospheric oxygen (Fig. 8.18, small image). Its "lungs" are thus located directly in its mouth and

Fig. 8.18 The electric eel emits short bursts of more than 600 V to catch prey and to defend itself. Its "lungs" with numerous well perfused papillae are located on the roof of its palate

supply it with the remaining 80% oxygen every 10–15 min. Its habitat is muddy white-water oxbow lakes, which are often low in oxygen. However, it is also found in clear water. For reproduction, the male builds nest of water plants, the eggs and the young fish are guarded by both parents.

All body organs including intestinal outlet are stowed directly behind the head in an approximately 10-cm-long section up to the gills. Following this up to the tail are densely packed converted cells of the fin muscles as electrical organs. These electrical cells, which comprise over 70% of its body volume, are arranged in rows and produce—each separately—a low voltage. In total, these up to 6000 electric cells deliver an active current pulse of more than 600 V (in large electric eels more than 800 V) at a current of one ampere over an extremely short period of only 2 ms. A simultaneous discharge of all electric cells is achieved by touching the victim or prey with the tip of the head (positive pole) or the tip of the tail (negative pole).

Electric eels generate surges for defense and hunting, but also for orientation in murky water and territorial defense. It is reported that they can paralyze a frog at a distance of one metre with light electric shocks. Larger animals are also usually stunned by its electric shock, allowing them to drown afterwards. This had already been impressively demonstrated by Alexander von Humboldt in 1800 with an experiment, as his diary notes show: "Our guide brought horses and mules and chased them into the water. Before 5 min had passed, two horses had drowned. The eel, six feet long, thrust itself against the horse's belly and gave it a blow."

I once had direct contact with an electric eel: with our tow net we had caught an electric eel in shallow water during the dry season. I tried to get the fish out of the large dragnet with my landing net, which had a metal frame with a metal handle. In this process, I felt a strong electric shock that had discharged when the landing net made contact with the fish. The electric current of the electric eel is enough to fatally injure even humans.

My good friend Moaçir Fortes Pereira, who has been roaming the Amazon rainforest with me for many decades, also told me that electric eels even make açai palms–standing on the river bank drop their tasty berries so that the fish can eat them.

8.2.7 Arapaima *(Arapaima gigas)*: Bony Tongue Family (Osteoglossidae)

Other name: Pirarucu (brazilian)

The species name is borrowed from an Indian language. Adults have a streamlined olive-green body with a seam-like dorsal and anal fin and a body length of well over three metres, weighing more than 200 kg (Fig. 8.19). This makes these living fossils, which have existed for 150 million years, among the largest freshwater fishes on our planet (along with the wels catfish—or

Fig. 8.19 One of the largest freshwater fish in the world is the Arapaima (Pirarucu), with a body length of more than 3 m

sheatfish—in central Europe). The largest specimens ever caught are said to have reached more than 4 m.

Because of this size and its lifestyle in warm, low-oxygen white-water floodwaters, it is an obligate air-breather, regularly coming to the surface every 5–15 min, where it inhales with a loud sound. Its swim bladder features lung-like structures with a large blood supply that can extract oxygen from atmospheric air through a capillary system. If they did not reach the water surface, they would drown.

Younger arapaimas like to jump out of the water like dolphins and then slap their 100 kg bodies on the water. They live carnivorously and feed on fish, frogs, waterfowl and small mammals. The way this works is that when the fleshy, drooping lips are torn open, a powerful suction is exerted to suck in the prey; when this happens on the surface of the water, there is a whip-like crack. In captive arapaimas I frequently observed this impressive spectacle near Manaus: After throwing the food fish into the water, the pirarucus came to the surface with a mighty flick of their tail and sucked in the prey with a loud crack.

At mating time, the tips of the large scales turn orange-red, which has earned this fish species the name pirarucu—the red fish (from the *Tupi language pira* = fish and *urucu* = red) (Fig. 8.20). To reproduce, they use their powerful caudal fins to strike a large depression in the sandy soil, in which about 50,000 eggs are laid. It is the male which then performs brood care, but only a few young fish survive the first few days.

The large, overlapping scales protect these fish like a tank. If a piranha bite hits one part of the body, the force is distributed over the scales in the surrounding area. In the Amazon basin, souvenir shops sell these attractive, 5-cm scales of this fish (the orange-red has since turned black) as well as its ossified, 8-cm-long, toothed rasping tongue for little money. The tongue is often used as a grater in the household.

Arapaimas are popular food fish that can be found in any market. Their fillets are also sold rolled up as salted dried fish. Since 2001 they are protected all year round and may not be caught. They used to be of great economic importance. I have also observed that over the years the arapaimas on offer have become smaller and smaller, barely reaching a metre in length. Away from the towns and villages, I encounter *caboclos* every year who still—despite the ban—fish arapaimas in large numbers from the flooded lakes with set nets and then sell them from boats on the river banks. They are also speared, the fisherman just has to wait for them to surface and a large circle of water is formed. Meanwhile, there are the first successful attempts to breed Apapaimas in lakes.

Fig. 8.20 This giant pirarucu (= red fish) was photographed 30 years ago at the *Manaus* fish market

Figure 8.20 shows a huge Arapaima, I took the photo 30 years ago at the fish market in *Manaus*.

Tip

In some German aquariums (like in Berlin and Cologne) as well as in the United States, for example in the Milwaukee Zoo, the Dallas World Aquarium and the Smithsonian's Zoo/Washington, you can face Arapaimas with more than one metre body length "eye to eye" and you can observe the behavior of these primitive fish.

8.3 Frogs

Due to the huge water areas, it seems understandable that more than 400 different amphibian species can be found in Amazonia. For comparison: In Germany, there are 19 species, worldwide more than 7000 species. With its constant temperatures and high humidity, the Amazon region offers ideal living conditions for frogs, which have an unprotected body covered only with moist skin. To protect them from pathogenic bacteria and fungi, their mucous membrane contains chemical components that usually keep such pathogens away.

Frogs are difficult to spot in nature, let alone photograph, because most of them are nocturnal or crepuscular, hiding in the shady undergrowth during the day. They are perfectly adapted to their habitat and are masters of camouflage, as Fig. 8.21 shows: At first glance, only green algae, grey lichens and grey-brown bark flakes can be seen in the photo. Only on closer inspection the observer can identify the well-hidden frog (a tree frog of the genus *Dendrosophus,* which belongs to the tree frog family (*Hylidae*)), whose colouration corresponds exactly to its environment.

On the other hand, frogs are all the more audible if you go by boat to the white-water tributaries of the Amazon or the white-water lakes shortly before nightfall, around 6:30 p.m. local time. It is the males that mark their territory with their calls and attract the females to mate. The frog concert is varied and almost deafening near floating aquatic plants; but only in white water, where they grow in large quantities and insects are in great abundance. A polyphonic

Fig. 8.21 Frogs are masters of camouflage. Here The tree frog in this photo is hardly recognizable

chorus of croaks, whistles, hammers, and trills can be made out. Many frog species have sound bubbles that amplify their calls like a sounding board. I have often experienced frog species that start croaking violently even during the day about 30 min before the onset of a thunderstorm accompanied by heavy rain—apparently in "anticipation" of the coming rain and the possibility of copulation.

For my shots shown in this book, I caught an interesting frog with the landing net at night, photographed it the next morning and released it back into the wild. It also happened often that the frogs jumped into the boat during the ride.

8.3.1 Tree Frogs (Hylidae) and Other Families

There are approximately 100 different species of tree frogs of the genus *Boana* (previously placed in the genus *Hypsiboas*) in Amazonia. They all live near the river bank on trees. Their fingertips are always enlarged plate-like as adhesive discs. They are all nocturnal and eat mainly insects, from mosquitoes to moths.

The **dotted treefrog** *(Boanas*—formerly Hypsiboas *punctata*; other name *polka-dot treefrog)* is about 4 cm in size, has smooth, pale green skin with reddish spots and an equally reddish, interrupted, narrow stripe down the side. Its large protruding eyes are white or black with a transverse black pupil (Fig. 8.22). Under UV light, the body glows a poisonous green. To reproduce, the female lays a gelatinous spawn ball containing about 15 eggs on the calm water surface, where the spawn floats up. The male remains near the clutch until the tadpoles hatch.

The **blacksmith treefrog** (*Boana*, formerly *Hypsiboas-faber* wears this name because its calls sound like hammer blows on an anvil. The females grow up to 10 cm, the males remain slightly smaller. Their skin has a chocolate-like ground colouration (Fig. 8.23). The males dig pits at the edge of pools, which they surround with a dam of mud. Numerous eggs are laid in this nest.

The **rusty treefrog** (also called giant gladiator tree frog, *Boana*, formerly *Hypsiboas boans*) also creates mud nests. The female grows up to 14 cm. It is nocturnal.

The **Bokermann's tree frog** *(Bokermannohyla luctuosa)*, on the other hand, grows to only 10 cm. Its smooth skin is pale green; its protruding large eyes are striking. In case of danger, it produces a stinking secretion. Nothing is known about its spawning behaviour.

Fig. 8.22 The dotted treefrog is nocturnal, as its large eyes reveal

8.3.2 Tarsier Leaf Frog *(Phyllomedusa tarsius)*: Subfamily of the Treefrogs (Phyllomedusidae)

This poison-green coloured, nocturnal frog, up to 7 cm in size, has a long-limbed shape. Its fingers have conspicuous adhesive discs (Fig. 8.24). It can oppose its fingers and toes, which allows it to grasp and climb up thin stems, which, incidentally, it does very quickly. Seeing its bright green body colouration, it is easy to imagine how perfectly camouflaged this frog is. However, one specimen jumped into our boat and could therefore be photographed later. Eggs are laid on a leaf above the surface of the water, the leaf being formed into a bag. The gelatinous mucus holds the bag together, which is open at the top and bottom. This allows the tadpoles to fall through the lower opening into the water when they hatch.

The species name is borrowed from the tiny, nocturnal *tarsier,* called *Tarsius* in Sulawesi. By the way, this frog species belonged formerly to the treefrog family (*Hylidae*).

Fig. 8.23 The calls of the blacksmith treefrog sound like hammer blows on an anvil

8.3.3 Buckley's Slender-Legged Tree Frog *(Osteocephalus buckleyi)*: Tree Frog Family (Hylidae)

This 5-cm brown frog has a warty, tubercle-like skin: Numerous bumps with nodular swellings sit on the surface of the body. The outside of the legs is covered with fringes of skin. The fingertips are covered with plate-like adhesive plaques, and the middle finger is greatly elongated. It is nocturnal and inhabits small waterways such as *igarapés*, also climbs trees and eats grasshoppers, moths, spiders and beetles. During the day it hides, for example under roots. The females lay up to 1000 eggs in vegetation-rich bank sections. Its tadpoles are poisonous to fish.

Fig. 8.24 The tarsier leaf frog is perfectly adapted in colour to its environment

8.3.4 South American Bullfrog *(Leptodactylus pentadactylus)*: Southern Frog Family (Leptodctylidae)

Other name: Smoky jungle frog

This frog belongs with a body length of the males of 20 cm to the giants among the South American frogs . During the mating season, its extremities turn orange-red, and its body is green-brown in colour (Fig. 8.25). Bullfrogs live near rivers and hide under foliage during the day because they are nocturnal. They are omnivorous and eat other frogs, lizards, snakes, even small birds and bats. When the prey is large, they press on the palate with their eyeballs and push the prey into the gullet. As a defense, they hold a poisonous skin secretion, and moreover they can inflate their lungs when in danger, causing the enlarged body to stand up on its four extremities. When caught, they emit a shrill whistling sound.

In the male, the thumb callus has developed into a hard spine. Together with a two-humped horn on the chest, it allows him to perfectly hold the female like a nutcracker. Their mating calls are loud and sound like cattle bellowing from a distance. The male builds foam nests in which the female lays about 1000 eggs. The tadpoles are very large, up to 8 cm, they are omnivorous and also eat their conspecifics. The metamorphosis is unusually short and the frogs hatch after only 4 weeks. Among the local population, this frog species

Fig. 8.25 Bullfrogs inflate their lungs in danger, enlarging their bodies and standing up on their four limbs

is on the menu; more often I have seen *Caboclos* specifically hunting these animals.

This group of frogs is found only in tropical regions of the southern hemisphere. Previously, they were included in the family of whistling frogs.

8.4 Reptiles

The vertebrate class of reptiles includes very different groups such as lizards, iguanas, snakes, caimans or even turtles, all of which occur in the Amazon lowlands. Thus, the large number of 378 reptile species recorded so far in Amazonia is understandable, whereas in Germany there are only 13. They live mainly on the banks of rivers and lakes as well as in the floodplain forests, whereby their diversity in blackwater regions is significantly lower due to the smaller food supply.

Reptiles are also masters of camouflage and many are crepuscular or nocturnal. So the traveler does not often get to see them. Nevertheless, there are always surprises when suddenly the visitors discover a lizard on a branch from the canoe or directly in front of the boat a snake swims from one shore to the other. Caimans can also be seen more frequently in the less populated regions, where they rest on sandbanks during the dry season, even in daylight. Turtles are very endangered in Amazonia, as hardly any *caboclo* spurns them when he sees one. Alexander von Humboldt described as late as the 1800s that huge quantities of turtle meat and eggs were offered in all the markets. Today, we encounter turtles in or near the river extremely rarely. A few times we bought turtles from the river people, linked with the hint that we will not eat them, but release them, as they are protected and threatened with extinction.

The probability of seeing reptiles is greatest in the dark. If you canoe into the small *igarapés*, oxbow lakes, floodplains and lakes at dusk—around 6:30 pm—you are sure to spot caimans and often snakes. The trick is to use spotlights to illuminate the shore and trees near the shore. The reflective white and reddish lights of their eye background reveal where reptiles (and other animals) are.

8.4.1 Green Ameiva Lizard *(Ameiva ameiva)*: (Teiidae)

Other names: Giant Ameiva, Amazon racerunner

The lizards include about half of all reptiles of the present time, which is about 3000 species. In the Amazon basin dozens of species are living that are rarely seen. The species described here, however, I have seen more often. It is sun-loving and diurnal, so it can be seen sunbathing in the flooded forests on tree trunks and on branches. It is very slender and grows to 40 cm in length, with the tail taking up half of the body length. Their basic body colouration is brown, but green patches of colour are always visible, as on the posterior dorsum, on the flanks, and the tail is also green in adult *ameives* (Fig. 8.26). They hunt insect larvae, beetles, spiders, cockroaches and grasshoppers.

The members of this family are distinguished by large rectangular scales that form distinct transverse rows ventrally and generally small granular scales dorsally. They have head scales that are separate from the skull bones.

8.4.2 Green *iguana (Iguana iguana)*: Iguanas (Iguanidae)

This large iguana can be seen more often during the day when you search the trees near the river banks with binoculars, even on the Amazon itself. It prefers dense vegetation with light and shade places, especially on trees whose branches reach over the water. In case of danger, for example, by a loud boat or even calls, it drops into the water from a great height. It dives and swims underwater into herbaceous bank sections, where it is well camouflaged.

Fig. 8.26 The green ameiva lizard hunts insects and spiders

Green iguanas are loyal to their location and defend their territory against other iguanas. Thereby it comes to mock fights (comment fights), where the head is nodded and the wattles are inflated. The advantage of these ritualized fights is that the risk of injury is low.

The green iguana reaches a total length of about 1.5 m, with the tail accounting for two-thirds of this. The male is significantly larger than the female. On its massive head sits a larger wattle. Further, he has enlarged gray-green scales sitting below the eardrum. Finally, it has a much larger dorsal crest than the females (Fig. 8.27). The body colouration of adults tends to be gray-green, the older they get the grayer they become, while young iguanas are coloured poison-green and thus even better camouflaged.

Iguanas feed purely vegetarian on leaves, flowers and fruits. To accelerate the digestion of hard leaves, they go into the sun to warm up, because a higher body temperature also intensifies its digestion.

Copulation takes place on trees. The male fixes the female during mating with a neck bite. The approximately 30 eggs are laid in a self-dug burrow, in which the young hatch after 30 days. There is no brood care.

River residents like iguanas, as they are very popular as meat-food. The locals call them with the nickname "tree-chickens".

Fig. 8.27 The male of the green iguana adorns himself with a large dorsal crest

8.4.3 Spectacled Caiman *(Caiman crocodilus)*: Alligator and Caiman Family (Alligatoridae)

The crocodilians include crocodiles (Africa, Australia), gavials (Asia), alligators (Florida) and caimans which exist only in the Amazon lowlands. Of the four species of caiman that occur in the Amazon region is the northern spectacled *caiman* also called common or white caiman. A bony transverse bar between the corners of the eyes, which resembles a pair of spectacles, helped it to get its English name.

Its maximum body length is given as 3 m (Fig. 8.28). The broad head, in contrast to crocodiles, is striking, and its jaws are densely set with pointed teeth. When the mouth is closed, *no* teeth of the lower jaw are visible in caimans, but in crocodilians, where the fourth tooth of the lower jaw is visible even when the mouth is closed; it lies in a laterally open groove of the upper

Fig. 8.28 A bony transverse bar between the corners of the eyes helped give the spectacled caiman its name. The small illustration shows a young spectacled caiman

jaw. Around its mouth are hundreds of tiny, extremely sensitive sensory cells called tactile mounds. They contain many sensitive nerve endings. Even the slightest touch of a tactile mound, for example in an experiment with a hair, triggers an immediate reaction: The animals snap to attention within a tenth of a second.

Spectacled caimans inhabit calm rivers, lakes and swamps, where they eat mainly fish, but also other reptiles, frogs and even birds. They are nocturnal; during the day they rest hidden in riparian vegetation or on the sandy banks overlooking the water. Sometimes I have seen them floating on the surface of calm water during the day.

Male caimans live territorially and have a hierarchy among themselves. With the onset of the rainy season, they build nests of earth and rotten leaves between the riparian vegetation, in which up to 40 eggs are laid. After about 70 days the young are ready to hatch, which they communicate to their brood-caring mother by softly grunting "Umf-Umf-Umf". Then the mother opens the clutch, after which the young burst the elastic egg skin with a hump on the tip of their snout.

Interestingly, crocodiles do not have sex chromosomes. In humans, the female sex is determined by the XX chromosome, the male by the XY chromosome. In contrast, crocodile sex is determined by external circumstances:

At egg temperatures below 32 °C, females are born, and at those above 32 °C, males are born. This means for the eggs, for example, that in years with consistently higher temperatures, only males are born. If due to the looming climate change, the temperatures in the Amazon also increase significantly permanently, almost only male caimans will hatch, so that this species could be threatened with extinction.

Hatching caimans are about 10–15 cm in size and have a yellowish base colouration with dark cross bands. They are protected by their mother for about 14 days, yet a large number of them do not reach the first year of life. They eat insects and frogs in the first year of life and reach a body length of 30 cm after 1 year. For 8 years they grow about 30 cm every year, then much slower.

If you canoe into calm white-water igarapés with weedy banks in the dark and use good headlights to illuminate the river banks, the reddish reflective eyes of the caimans give away their location. What glows here is the *tapetum lucidum* (= "luminous carpet"), a thin reflective layer of skin behind the retina at the back edge of the eye. This layer enables the caiman—and, for that matter, nocturnal birds, snakes or even cats—to perceive prey despite the low light. The little light penetrates the retina, is reflected by the *tapetum lucidum* and reaches the retina again within fractions of a second so that the amount of light is practically doubled and the caiman gets to see its prey.

If the eyes are close together, it is a young caiman and you can try to get it out of the water for a short time to take a photo. However, it depends on the skill of the boat driver and the catcher to come to a stop in front of the caiman, if possible without big waves and noises, so that the catcher can grab the caiman behind the head before it dives. Caught by the hand the caiman remains completely motionless, because the momentary stress causes the release of hormones that paralyze its muscles for a short time. But be careful: you never know how long the immobility will last. A good grip on the animal's neck will prevent it from suddenly running around in the boat.

Caimans can swim up to 30 km per hour, especially since their short acceleration is enormous. After each dive, they come back to the surface after 4–6 min to catch their breath, but they can also stay underwater for more than 15 min.

In addition to the spectacled caiman, the black caiman *(Melanosuchus niger)*, which is darkly coloured on its back and sides and grows to a length of well over four metres, is also more frequently seen, especially in black water. Unlike the spectacled caiman, its population is highly endangered, as it is still illegally hunted to obtain its leatherskin (Fig. 8.28). Spectacled caimans, on the other hand, can still be seen at night times in the Amazon basin. Pretty often, although their numbers have also declined sharply because river people also eat their meat. As late as the nineteenth century, British naturalist Henry Walter Bates wrote: “The waters are as filled with caimans in the dry season as the pools in Europe are with tadpoles.”

By the way: For every 100 people bitten by snakes, one dies. In contrast, for every 100 people bitten by crocodiles or caimans, 13 die—mostly from an infection due to blood poisoning, because the teeth are filled with germs that are life-threatening for us. Anyone bitten by a caiman should therefore seek medical treatment within six hours, otherwise, there is a risk of blood poisoning and amputation of the affected limbs before life itself is put at risk.

8.4.4 Anaconda and Other Boas—Boas (Boidae)

Other name: Sucuri (brazilian)

Of the about 60 boa species, the **green anaconda (*Eunectes murinus*)** is the largest, with a recorded body length of 9 m, a weight of over 200 kg, and a girth of over 1 m when a large prey has been devoured. It is considered the largest snake in the world. Time and again, there are reports of even larger specimens. Therefore, the *National Geographic Society* has offered a prize of US$10,000 to the person who can present a photo of an anaconda that is at least 10 m long.

Their basic body colouration is olive-green to light brown with the characteristic oval black spots (“mirrors”) on the back (Fig. 8.29a).

The lidless eyes and nostrils sit conspicuously high on the head, indicating their aquatic lifestyle. The females are significantly larger than the males. They inhabit slow-moving or still waters of all three water types (white, black, and clear) with dense riparian vegetation. They can dive for up to 30 min and also like to lie underwater in air pockets between tree roots for long periods, waiting as ambush hunters for their chance, for example, a fish swimming by. Their range of prey also includes frogs, reptiles—even 2-m-long caimans—birds and mammals. With their deeply forked tongues, anacondas sniff out even the slightest scent of their prey (Fig. 8.29b).

All boas are non-poisonous, but their teeth are very sharp and can also contain dangerous germs, which can also be life-threatening for humans if they bite. Similarly, all boas belong to the strangler snakes, which means that they do not kill their prey with venom, but entwine them with several loops of their muscular body. With each exhalation of their victim, the ring becomes tighter and tighter, so that eventually important blood vessels burst and the

Fig. 8.29 (**a**) With a body length of up to 9 m, the anaconda is the largest snake in the world, (**b**) With their deeply split tongues, anacondas sniff out even the slightest scent of their prey

circulation collapses and the prey dies. For a large prey—such as a caiman—it can open its upper and lower jaws up to 180 degrees with the help of highly elastic bands. To prevent the victim from slipping out of its jaws in its death throes, the upper jaw contains four rows of 100 very sharp teeth bent backward. The gastric juice of the anaconda contains—as with other boas—a highly concentrated hydrochloric acid solution, which also enables bone digestion. Only teeth, hair and claws are excreted again. Green anacondas need larger prey about every 50 days.

Every 2 years, the female anaconda is ready to mate. For this purpose, a true orgy of copulation takes place. Dozens of petite male suitors and the much larger female entwine and form a huge mating tangle. Who finally carries out the actual fertilization cannot be determined, but the female copulates with several males in succession. After about 7 months, during which she fasts and loses 20% of her weight, up to 100 fully developed young are born alive; they are then already 80 cm long and can immediately go hunting for prey on their own. Nevertheless, many fall victim to their predators, such as birds of prey or caimans, in the first few weeks.

I have only seen green anacondas a few times during my trips, as they are very cautious and usually only their head can be seen on the water surface between riparian plants. Once, during a night excursion, I was lucky enough to see a large green anaconda for a few minutes swimming right next to the side of my canoe in a fast-flowing *Igarapé*. It was so large that I could only capture its head with my telephoto lens in the headlights (Fig. 8.29b).

The second-largest strangler in the Amazon basin is the red-tailed boa *(Boa constrictor).* family*: Boas*), which can be seen with a bit of luck if you travel from the flooded lakes or rivers into the flooded forests during the rainy season. It is important to use binoculars to search the thick branches from the boat, as the *boa constrictor* likes to rest on these during the day. You can also ask in the villages on the river banks if anyone keeps a red-tailed boa at home, as they like to keep them as pets in a wooden hut and feed them with trapped mice and rats. Several times I had success with this and was able to photograph the snake.

This snake grows up to 4 m long with a stately 27 kg. Its basic colouration is mostly reddish-brown with dark-framed light spots on the back, so-called "saddles". The tip of the tail is a particularly contrasting reddish-brown (Fig. 8.30).

It always lives near water in dense vegetation, during the day it sleeps in hollow trees or in the shade of its crowns on thick branches. With the onset of dusk, it becomes active. It eats all animals that it can overpower in size—either as a lurking hunter or actively, following the scent trails of the prey animal. The feeding pattern is similar to that of the anaconda: snapping and then crushing and devouring the victim. Recently I was able to convince myself of the muscular strength of a still young *boa constrictor* on my own body. In a village I took the snake, which was about 1.20 m long, out of its cage, holding it with my hand behind its head. Thereupon it entwined my arm with several rings of its entire body, pulsatingly exerting enormous pressure on my arm with its body muscles about every 2 min. Yet her body was

Fig. 8.30 The *boa constrictor* is a lurking hunter

only about four inches in diameter. What power has then a full-grown *boa constrictor*; with a body diameter of over 20 cm!

The species name *constrictor* is also very aptly derived from the Latin verb *constringe* (= to strangle).

Once we surprised a red-tailed boa in the jungle shortly after devouring its victim. Since it was disturbed during digestion, it choked out the completely slimed victim again shortly afterwards. It was an 80 cm long opossum.

The riparian rainforest is full of surprises—like the colourful **rainbow boa** (***Epicrates cenchria***; family: Boas). It lives on the ground near the water between low vegetation. So you may meet it during a forest excursion. It grows to a maximum length of 2.3 m, in which case its body is just under arm's length. Its basic colouration is coppery red-brownish with black circles as a pattern. Its entire body is covered with iridescent scales that produce rainbow-like reflection patterns when sunlight strikes it (Fig. 8.31).

It has heat-sensitive sensory cells along its lips. This allows it to follow its prey. It also kills its victims such as lizards, frogs and birds by wrapping itself around them and strangling them until the victim goes into cardiovascular failure.

Fig. 8.31 The rainbow boa is covered with iridescent scales that give rise to rainbow-like reflex patterns

Finally, the **Amazon tree boa**, also known as **garden tree boa** (***Corallus hortulanus*** family: boas), should be mentioned here, as it is very often found in the trees near the river bank, where it hides during the day and is impossible for us to identify. But in the darkness, there is a good chance to spot them in the branches with the help of a spotlight. So I have often been able to observe these snakes. They are slender in shape and grow up to 1.5 m long. Their wedge-shaped, clearly separated head remotely resembles a dog's head, hence their German name (Fig. 8.32).

Its species name *hortulanus* indicates that it was observed in a garden when first described.

Their body colouration shows an immense variety of colours and patterns anywhere from black, brown, gray to orange, yellow or many colours in between. It may or may not also have distinct spots on its back. The male has a thorn-like spur under his tail, which he uses for defense and copulation. These boas are viviparous.

They are night hunters, their prey consists of small birds, frogs, lizards and small mammals. The deeply recurved, needle-like fangs can cause deep wounds. Once an Amazon tree boa bit me in the palm when I caught it with the landing net and wanted to hold it in my hand for better photography. I guess I had stressed it too much ... This snake species has two biting techniques: a mild or gentle bite for defense and a deep, aggressive bite for catching and later killing its prey. In my case, fortunately, she had only used her

Fig. 8.32 The Amazon tree boa is a night hunter with deeply recurved, needle-shaped fangs

gentle bite, which only left an imprint of her teeth in my palm, which then filled with blood, but later had no negative health consequences.

8.4.5 Mata-Mata *(Chelus fimbriatus)*: (Chelidae)

This large aquatic turtle with a body length of up to 40 cm lives in calm, predominantly black-water rivers, lakes, stagnant pools, marshes and swamps. It is striking that the head and neck are almost as long as the rest of the body.

In the water, especially since it is extremely well camouflaged, it can only be seen by us with great luck. But from time to time you can see them on the canoes of the *Caboclos*, who catch them with nets and eat them.

The photo shown here was taken after we had bought a matamata—as it is called in the *Tupi language*—from a river riparian and then released it into the wild (Fig. 8.33).

Pyramid-shaped bulging shields sit atop its flat and dark brown dorsal carapace. At the top of the head sits a thin snorkel-like proboscis, which she can use to catch her breath without surfacing. Its tiny eyes also sit far forward on its head. There are protruding, fringed flaps of skin on the underside of its head, neck and side, which camouflage it perfectly. Between its toes, it has webbed feet, which enable it to swim quickly. Sharp claws on its toes make it defensible. At the bottom of a stagnant pool, for example, it lurks for its prey. These are mainly small fish, which are simply sucked in. It reproduces between

Fig. 8.33 Pyramidal horny shields on the dorsal carapace of the mata-mata

October and December in the upper Amazon region by laying about 20 eggs in a nest hollow made of mud.

8.5 Birds

Already when driving from the airport to the hotel in Amazonia, for example in *Manaus,* the visitor notices birds. Black vultures are circling in the sky, swallows darting through the air and parrots roaring in the dense treetops at the road junctions. Birds are omnipresent in the Amazon lowlands: there are over 2000 species in this region, which is still about 20% of the approximately 10,000 species identified worldwide. One reason for this enormous abundance is the huge amount of food sources for birds such as flowers with their pollen and nectar, fruits and seeds, insects, fish and for birds of prey the many bird species themselves. It is therefore worthwhile to have a pair of powerful and compact binoculars with you on your trip to the Amazon.

When visiting the Amazon region, we see birds in different habitats—in urban zones like cities, in open areas, in the rainforest and of course near and on the rivers and lakes. Of course, there is also overlaps in the distribution of birds. For example, many forest-dwelling bird species or urban birds can often be seen on riverbanks as well. Therefore, the birds described in this book will be divided into only two main habitats: "Birds on the water" and "Birds in the rainforest". In the selection of species for this book, the same guidelines do apply as for the two previous sections: what is presented is what the visitor can really see. That is why such unusual bird species as the *Harpy Eagle* with its 2-m wingspan or the boat-billed heron (boatbill) remain undescribed here, as one hardly ever gets to see them.

The following sections show the incredible perfection with which birds have learned to survive in their habitat over the course of evolution, and the astonishing and diverse strategies they have developed to feed themselves and maintain their species.

8.5.1 Large-Billed Tern *(Phaetusa simplex)*: Terns (Laridae)

Surprisingly, there are no gulls in the Amazon river system, this habitat was rather occupied by terns. The species described here is ubiquitous especially in the main stream, but also in its tributaries and in the rainy season in the huge *várzeas*. Striking features of the large-billed tern, which grows to 40 cm, are its namesake enormous yellow bill and black head cap (Fig. 8.34).

They like to form smaller flocks and sit on logs floating in the river. To find food, they shake and then dive for fish from the air. Like the gulls, they are

Fig. 8.34 The large-billed tern is ubiquitous in the Amazon river system

opportunists and also grab the leftovers from the fish market in the villages along the river.

8.5.2 Wattled Jacana *(Jacana jacana)*: Waders (Jacanidae)

This species inhabits water surfaces of floodplain meadows overgrown with aquatic plants—such as the Queen Victoria's water lily *(Victoria amazonica)*, water cabbage (*Pistia)*, eared moss *(Salvinia)* and *Eichhornia*. On these swaying leaf surfaces, they find optimal footing with their oversized feet with extremely extended toes. The distance between the front and back ends of their toes is 23 cm—which is as long as the entire size of these birds. Their wings are reddish-brown in colour, their head, neck and breast are black, their face and frontal lobes are red and their bill is yellow (Fig. 8.35).

Between the floating aquatic plants, they search for insects, crustaceans and snails. They are poor fliers, flying up in danger with a loud clamor and usually landing on the plants again after only a short distance. When they fly up, their light yellow-green wing undersides become visible, which has earned them the nickname "butterfly birds".

For reproduction, the dominant female mates with several males. This then builds a nest on the huge leaf plates of the Queen Victoria's water lily at the beginning of the rainy season, which mostly consists of the small leaves of the eared moss and of the floating *ludwigia*. The four beige and dark brown

Fig. 8.35 The wattled jacana lives between the floating leaves of the *várzeas*

speckled eggs are perfectly camouflaged and cannot be seen even from a short distance. The protection of the nest and the raising of the young is also done by the male. The name *Jacana* is a proper name from the *Tupi language*.

8.5.3 Herons: (Ardeidae)

The various heron birds are most commonly found on the banks of the Amazon River system. Generally, they have long legs and a long neck compared to their overall body, as well as a dagger or spear-like bill. They prefer to stay in shallow water near the shore, where they wait motionless for their prey as ambush hunters and strike with lightning speed with their beaks. Their food spectrum is wide and includes fish, frogs, reptiles, small mammals, insects, crustaceans and eggs.

Most heron species are very shy and fly up already at an approach at a great distance. Many are sociable and also spend the night together—like the great egrets and parrots—on roost trees, which then look like after a snowstorm. In addition to protection from predators, such roosting trees have the function of *chat rooms*, for example by exchanging information about waters rich in fish or such trees that just provide ripe fruit in large quantities.

For reproduction, herons establish colonies. First, the male birds emerge with now conspicuously elongated decorative feathers on their heads and necks. They then perform ritualized head and wing movements before copulation occurs. The nests are built jointly by both sexes and consist simply of an accumulation of scrawny twigs. The clutch consists of three to five eggs, the brood care is taken over by the two adults together.

The most common heron is the **great egret** (*Ardea alba*), which grows up to one metre. Its size and the colouring of its bill and feet are its distinguishing features from the snowy heron (see below). In the great white egret, the bill is yellow and its legs and feet are black (Fig. 8.36a).

In the case of the **Snowy egret** *(Egretta thula)*, it is the other way round: black bill and yellow feet (Fig. 8.36b). It is also much smaller at 60 cm. While the great egret waits motionlessly for its chance to spot a fish, the snowy egret moves its long, yellow toes back and forth in the water to attract prey (which supposedly want to snatch a worm) or also to track them down.

The third white heron is the **western castle egret** *(Bubulcus ibis)*, which at around 50 cm is the smallest of the three, its bill and feet being yellow. It is less common on the banks of Amazonian rivers but is more likely to be found in the lower reaches of the Amazon on cleared areas with cattle pastures.

Fig. 8.36 (a) In the great egret—here in courtship dress—the bill is yellow, and the legs and feet are black, (b) In the snowy egret it is the other way round: black bill and yellow feet

Fig. 8.37 At 1.30 m, the cocoi heron is the largest heron in the Amazon lowlands

The **cocoi heron** (*Ardea cocoi)*, with body size of 1.30 m, is the largest heron of the Amazon lowlands. Its plumage is gray-white-black with a black cap on the top of the head that extends to the eyes. Its bill is orange in colour (Fig. 8.37). This species of heron usually nests singly in trees, protected from predators, in marshes. Its species name *cocoi* is probably derived from Chilean, where this heron is called "Cuca" after its typical call.

The 60 cm **capped heron** (*Pilherodius pileatus*) is unmistakable. Its neck is reddish brown-cream coloured, the top of its head has a black cap. The face and base of the bill are blue. Long, white decorative feathers are on the back of its head (Fig. 8.38). It is active at dusk. For reproduction, the male performs elaborate courtship flights combined with wing spreading. The adults then build a nest together from twigs in the trees. The young birds are cared for several weeks.

The **yellow-crowned night heron** (*Nyctynassa violacea*) is one of the two species of night herons with a body size of just under 60 cm and resembles the other one being the **black-capped night heron** (*Nycticorax nycticorax*) in appearance. It is grey in colour with mottled black and grey upper wings. Its head is black and glossy, white cheeks and a pale yellow crown during the mating season.

There is a white stripe under the eyes. Its eyes are amber. During the mating season, it wears additionally long white decorative feathers on the back of its head. It nests in trees near the shore, where it builds a nest of loose twigs. The

Fig. 8.38 The capped heron is unmistakable with its long white decorative feathers on the head and blue face

black-capped night heron shows besides the black crown and black back red eyes. Two or three long white plumes, are remarkable, when erected in greeting and courtship displays. They extend from the head. Since both species are also found in the Caribbean, where they specialize in hunting crustaceans at night.

The **rufescent tiger heron** (*Tigrisoma lineatum*) is a larger heron with a body size of up to 90 cm. It is very strikingly coloured to the human eye with its bright maroon head and neck. Its back is black-green with a very fine spotted pattern. Its bill is ten inches long. The iris is yellow (Fig. 8.39). When threatened, it assumes a bittern-like position—similar to a pole, which camouflages it optimally *(somatolysis).* As a solitary bird, it prefers locations close to the shore with dense undergrowth, where it lies in wait for its prey at night. Besides fish, it likes to eat snakes and dragonfly larvae. The young birds look completely different and have a tiger-like feather pattern (Fig. 8.40). That is why *Tigrisoma* (= tiger body) has been chosen as its generic name.

The **mangrove heron** (*Butorides striata*), also known as striated heron or green-backed heron, is one of the small herons with a body length of 40 cm.

Fig. 8.39 The young birds of the rufescent tiger herons have tiger-like feather markings

Its plumage is grey-green with a black cap. Its bill is also black (Fig. 8.41). It lives hidden on riverbanks and lakeshores, where it lies in wait for its prey as a loner, even at night. It uses a trick to do this: It deliberately places insects as bait on the water surface to attract fish. When grabbing the prey, its head moves forward in a flash, stretching its neck very long.

For reproduction, both adults build the nest in bushes whose branches hang over the water. Since its distribution also extends to the mangrove forests of the Caribbean, it bears this name.

Fig. 8.40 The adult rufescent tiger heron is very conspicuously coloured

8.5.4 Neotropic Cormorant *(Phalacrocorax brasilianus)*: Cormorant Family (Phalocrocoracidae)

These birds, up to 70 cm in size, are also called olivaceous cormorants because of their shiny metallic black and green plumage, especially during courtship. The base of the bill is bright yellow and the iris is green (Fig. 8.42). The young birds are light brown in colour. In the Amazon lowlands, they are often seen hunting on the rivers. Or they can be seen at dusk, flying in formation as if strung on a string of pearls, with rapid, powerful wing beats close above the water surface to their roosting trees. Their catching technique is interesting: they thrust torpedo-shaped into the water, dive and pursue their prey with powerful wing movements until they finally seize it with their strong beak. I have more often seen them capture catfish, which have razor-sharp fin rays. They then throw the prey up, again and again, twisting it until they finally swallow it head first (Fig. 8.42). They like to hunt in groups, pushing prey toward each other. Since they live without the fat-producing brush gland, the

Fig. 8.41 The mangrove heron places insects as bait on the water surface to attract prey fish

feathers must be dried after the dives, for which they like to spread their wings wide.

They breed in colonies. There are often hundreds of breeding pairs, building their simple nests from branches in trees. During the dry season I have observed several times that thousands of these cormorants gather on sandbanks, walking slowly in circles—an unbelievable spectacle that is still not clearly explained. Possibly the food supply in the shallows is so tremendous that the birds congregate there in large numbers in anticipation of easy prey.

Some species of the cormorant family are called "*shags*". This name goes back to sailors who called all cormorants so.

8.5.5 Darter *(Anhinga anhinga)*: Darter Family (Anhingidae)

Anhingas are closely related to cormorants, their body shape is similar, but they are up to 1 m in size. Their neck and breast part are light brown, from which the rest of the body is coloured black. The wing coverts are striped black and white. The long, spear-shaped bill is long. Its long, broadly fanned tail and greatly elongated neck are conspicuous (Fig. 8.43). When swimming, only this protrudes from the water, as their bones—unusually for birds—are

Fig. 8.42 The catfish is rotated in its beak until its razor-sharp fin rays cannot injure the olivaceous cormorant when it is swallowed down

hardly filled with air. This is why they are also called "snake birds". The name anhinga comes from the Tupi language meaning "devil bird". Their flight seems to be arduous: powerful wing beats always alternate with longer gliding phases. All toes are equipped with webbed feet, so they can dive perfectly, but only for a short time of about 1 min. During this time, they surprise their prey, which they literally stab with their long beak. Like cormorants, they dry their wet plumage in the sun by spreading their wings. They inhabit river banks with dense vegetation and branches protruding into the water, on which they can rest. They breed at the edge of cormorant colonies.

Fig. 8.43 The darter is a relative of the cormorants

8.5.6 Kingfishers: Family—Kingfishers (Alcedinidae)

There are five species of this bird family in the Amazon lowlands, three of which—the most common—look quite similar. However, they are very different in size, which facilitates their identification. The most common is the ringed kingfisher, which is the largest of these three species, up to 42 cm. All kingfishers live territorially on wooded shores of lakes or slow-moving waters. Their territorial boundaries are formed by prominent perches, for example, thick tree branches close to the river bank or branches projecting into the water, which they can easily approach. In case of danger—for example when a canoe approaches—the ringed kingfisher, in particular, makes a loud and very excited noise and flies to its nearest perch. When the boat approaches again, he flies up again and sits on the next perch until he finally reaches his territorial limit, from where he flies back. In order to photograph these birds well, you should keep an eye on the spot where he landed and then wait until the boat approaches this point.

The beak is spear-like long in all species. The way of life, diet and reproduction are similar in all species. The ringed kingfisher digs breeding tubes up to 2 m long in sandy quarries in the river bank area or cliffs, as in the middle reaches of the Amazon (Fig. 2.3). At its end, this tube widens to form a breeding cavity. Both adults care for up to four eggs and the young birds for about 5 weeks after hatching. The brood cavity is not cleaned of food remains and droppings.

In the **ringed kingfisher** (*Megaceryle torquata)* the back is rather grey, but in the male the entire breast is maroon or reddish-brown, with a white collar above. On the head, it may erect a shaggy crest (Fig. 8.44). In the female, the upper part of the breast is grey, with a narrow (sometimes incomplete) white ring joining the belly below; the rest of the belly is also reddish-brown. It

Fig. 8.44 The breast of the male ringed kingfisher is reddish-brown, the plumage otherwise rather grey (size: 42 cm)

Fig. 8.45 The male Amazon kingfisher has a reddish-brown breast and metallic green plumage (size: 30 cm)

grows up to 42 cm. The front toes are fused at the base and serve as shovels when digging the gross tunnels. The species name *torquata* translates as "collared" and refers to the white-collar on the male.

The **Amazon kingfisher** (*Chloroceryle amazona*) is about 25% smaller than the ringed kingfisher at 30 cm. Its plumage is distinctly metallic green in colour, with the nape and throat white (Fig. 8.45). In the male, the breast is chestnut brown, but in the female, it is white with green spots on the sides. In this species, there are no white markings on the wing as in the green kingfisher (Fig. 8.46).

The **green kingfisher** (*Chloroceryle americana)* is the smallest kingfisher in this trio at 20–25 cm. Its plumage is green with white spots on the wings. There is a white band around the neck. The male has a chestnut-brown breast (Fig. 8.46), whereas the female's breast and belly are dirty white.

8.5.7 Osprey *(Pandion haliaetus)*: (Pandionidae)

This quite small bird of prey of just under 70 cm has a wingspan of about 1.7 m. The head is white with a sharply defined dark brown eye stripe. The back and the upper side of the wings are dark brown, the tail is light-dark

Fig. 8.46 The breast of this male green kingfisher is not yet reddish-brown, the plumage is green with small white spots on the wings (Size: 20–25 cm)

Fig. 8.47 The osprey has a bright white coloured body underside

banded with a dark end band. Its predominantly bright white underside of the body with a brownish band on the breast is striking. The wings and tail are darkly banded below. Its long, pointed claws are strongly curved (Fig. 8.47).

Its habitat requirements are low: slow-flowing waters rich in fish. And there are plenty of these in the Amazon lowlands. When it has spotted a fish, it pushes into the water with its feet stretched out in front of it and grabs the prey.

It builds its large nests of branches on high trees. Due to its loyalty to the nesting site, it often mates again. The clutch consists of two to three eggs, the young birds fledge at just under 8 weeks.

8.5.8 Buzzards/Hawks: Family—(Accipitridae)

There are several species of these medium-sized birds of prey in the Amazon lowlands, as the food supply is very large. They like to stay close to the water, as they have a better view of their prey in the riparian areas than in the rainforest. From the boat, therefore, you will often see these birds circling in the sky. They are strikingly broad-winged and short-tailed hoverers. Their curved bill is short, and their feathers are often cross-banded on the underside of their bodies. Their prey range is wide and consists of insects, carrion, frogs, reptiles, birds and small mammals.

The **roadside hawk** (*Buteo magnirostris*) is one of the most common birds of prey in Amazonia, it even likes to stay in human settlements, it is not picky about its prey and is considered an opportunist. It is barely 40 cm tall, the male is even slightly smaller than most raptor species. Its breast is mottled brown and white, its tail has four to five horizontal stripes, and its iris is yellow (Fig. 8.48). In-flight it is unmistakable because the tail is conspicuously long and the wings unusually short.

Its species name *magnirostris* is composed of *magni* (= large) and *rostris* (= beak), as its beak is quite large for buzzards.

The **savannah hawk** (*Buteo meridionalis*) is a medium-sized bird of prey of about 60 cm. Its reddish-brown body colouring is particularly striking, making it easy to identify. Its breast, belly and neck are covered with narrow black and white speckles, its tail is distinctly banded in black and white, and its iris is amber (Fig. 8.49). It inhabits open dry areas (hence its name) and wide marshes, where it is always on the lookout for prey on high perches. Its diet includes small mammals, lizards, snakes, crustaceans, and insects. It likes to build its nests of twigs and grass in the fronds of palm trees. Its clutch consists of only one egg.

Fig. 8.48 The roadside hawk is only about 40 cm tall

Fig. 8.49 The savannah hawk is easily recognised by its reddish-brown feathering

Fig. 8.50 The black-collared hawk is unmistakable by its white head above the cinnamon-coloured body and the black half-ring on the neck

The **black-collared hawk** (*Busarellus nigricollis*) is also a medium-sized bird of prey of about 60 cm and a wingspan of 1,40 m. Its body colouring is cinnamon, with the head strikingly white in colour. Its mostly fanned black tail is short. The iris is reddish. A black half-ring is on its neck (Fig. 8.50). When it flies, you can see its black hand wings. Its toes are thorny-soled and end in strongly curved claws. It inhabits a wide variety of watercourse types, except that the flow must be slow. It prefers to perch on a prominent approach from where it observes its surroundings. When it has spotted a possible prey, it flies off, glides a short distance and then grabs the prey from the water. If its feathers get wet in the process, it has to spend a long time drying them afterwards, as its feathers are not water-repellent. To breed, it builds a large nest of twigs and leaves in tall trees near water. Its clutch consists of one to two eggs.

The **snail kite** *(Rostrhamus sociabilis)*—is a smaller bird of prey with a body size of 40 cm. Its plumage is a striking ash-blue-grey, from which the sunny yellow field of vision and the red iris stand out in contrast. The long legs are also strikingly yellow-orange. The root of the tail is white (Fig. 8.51). This harrier prefers to stay in the vast floodplain meadows, where it glides along over the large carpets of floating leaves, looking for apple snails *(Pomacea*

Fig. 8.51 The snail kite has distinctive ash-grey or grey-blue plumage. It is specialized on apple snails

urceus). The captured snail, the largest have a shell up to 11 cm in diameter, is either brought to a perch or it is immediately cracked on a large floating leaf of *Victoria amazonica*. For this purpose, the thin hooked beak is perfectly suited. The sphincter of the snail shell lid is severed, after which the fleshy snail can be pulled from its shell and eaten. I have observed that on several leaves of the Queen Victoria's water lily lay up to ten empty shells, here was therefore a sumptuous snail meal.

By the way, the *caboclos* also eat these snails. However, with undercooked snail meat parasitic worms can infect humans.

For reproduction, this hen harrier builds a 30 cm nest of water plants on branches near the shore. The clutch consists of two to three eggs.

8.5.9 Falcons and caracaras- Falconidae

The members of this family are birds of prey, scavengers and strangler at the same time—and thus omnivores. Their size varies between 40 and 45 cm.

They live in gallery forests of rivers, where they grab fish from the water or consume dead fish. In addition, as typical food opportunists, they also eat various fruits and forage for insects in livestock pastures. I even observed a yellow-headed caracara pulling insect larvae from the skin on the back of a Cebu cattle. Finally, they also hunt small birds. They are also known to be egg predators, grabbing nestlings as well. I witnessed a caracara clinging to the hanging prey nests with its claws, tearing a hole in the nest wall with its beak, then pulling out a kicking nestling and flying away with it in its beak. Little is known about the reproductive behaviour of these birds.

The **yellow-headed caracara** *(Milvago chimachima)* has a cream to beige plumage on the neck, breast, and belly, the wing coverts are blackish-brown, the eye area is yellow, and a black band extends behind the eyes. Its tail feathers are banded black and white (Fig. 8.52). It is one of the common birds in the Amazon lowlands, seen roaming over the rivers and lakes, especially during the morning and evening hours. Since they are food opportunists, as described above, they can be fed with fish waste, which they grab elegantly from the water surface (Fig. 8.52). Its genus name *cararaca* and its species name *chimachima* are Indian-Spanish onomatopoeic and refer to the buzzing calls of these birds.

Fig. 8.52 The yellow-headed caracara is a food opportunist

Fig. 8.53 The black caracara has glossy black plumage

A close relative of the yellow-headed caracara is the **black caracara** (*Daptrius ater*). It has glossy black plumage with an orange-red, sometimes yellow, head and orange legs (Fig. 8.53). In flight, a white under-plumage with narrow white horizontal stripes is visible at the root of the tail. It prefers the banks of rivers and lakes. Its prey range is wide and includes frogs, small lizards and birds as well as (dead) fish. Attacking birds' nests is also part of its normal behavioural repertoire. The species name of the black caracara "*ater*" means "black".

8.5.10 New World Vultures (Cathartidae)

The New World vultures include seven species that live on the American double continent. Of these, four are found in the Amazon lowlands. The birds, which are related to the storks, live mainly on the river banks, rarely in the rainforest itself. The wing feathers, which are not greased, are spread out to dry after the rain. The black vulture in particular is a hemerophile that follows humans and likes to live close to humans. All of these vultures eat mainly carrion. None builds nests, by the way!

The **black or American black vulture** (*Coragyps atratus*) grows to 70 cm in size with a wingspan of 1.5 m. It has glossy black plumage and its head and neck are featherless and covered with wrinkled grey skin (Fig. 8.54). Its short

Fig. 8.54 In black vultures, the short wings prevent elegant, aerodynamic flying

wings are conspicuous, preventing elegant aerodynamic flight and glide. Thus its flight pattern is characteristic and consists of a constantly alternating flutter and short glide.

Raven vultures are very gregarious, especially at their feeding grounds. They are widespread in the Amazon lowlands and are especially ubiquitous in human settlements, including larger cities. Since humans associate them with rubbish and dirt and always scare them away, they are quite shy. On a dead prey, the eyes and tongue are eaten first, then the anal region It has also been observed more frequently that they eat dog excrement and cattle dung. Outside of cities, they eat defenseless prey such as lizards and young birds, as well as insects and fruits, in addition to carrion. To breed, they lay the two eggs without any padding at the bottom of tree hollows or stumps. They are extremely cautious and stealthy when approaching their "nest" and also when flying away.

The **tropical turkey vulture** *(Cathartes aura ruficollis)* is a subspecies of *Cathartes aura.* Its small naked head without feathers is particularly bright red (Fig. 8.55). It is darker and more black than *C. aura*, which occurs rather in North American regions. Its long wings that enables this species to circle for long periods in tropical thermals. It grows up to 75 cm tall with a wingspan of 1.9 m. It prefers to live on riverbanks, occasionally in rainforests. Near human settlements, it eats dead animals. Its sense of smell is extremely

Fig. 8.55 The bright red, naked head gives the turkey vulture its name. In contrast, the yellow-headed vulture (small image) has a brightly coloured naked head, with the yellow parts of the face dominating

sensitive. Thus it flies gliding and circling at high altitude, picking up low scents of carcasses.

The genus name goes back to the Greek *cathartes*, which means "cleaner". The species name *aura* comes from the *Tupi language* and means "bird of prey", the subspecies name *ruficollis* means "red neck".

The **yellow-headed vulture** (*Cathartes urubutinga*) is rather small, measuring 60 cm and with a wingspan of 1.60 m. It has black plumage. Its plumage is black, its naked, featherless head yellow-orange, its forehead and nape red, and its crest blue-gray (Fig. 8.55). The legs are whitish-pink. This species is more rarely encountered in the Amazon lowlands. It prefers virgin forests and riverbanks away from human settlements. Equipped with an excellent sense of smell, they search for food by flying close to the ground. They can often be seen circling above the riverine forest.

King vultures (*Sarcoramphus papa*) are admittedly not common in Amazonia, but with a little luck, you can observe them when there is a large carcass. So I was able to observe five of these impressive vultures on a dead dolphin on a riverbank over a longer period of time. They are up to 85 cm tall with a wingspan of 2 m. Their impressive head colouring is unmistakable,

Fig. 8.56 King vultures have a prominent, wart-like bulging skin flap above the beak

ranging from black to red-blue and red to orange and yellow. Above the bill is a distinctive orange, warty, bulging flap of skin (Fig. 8.56). The iris is white.

They usually live in pairs and soar during the day at high altitudes above the rainforest, where they can circle for hours without moving their wings thanks to the constant thermals. With their good sense of smell and sharp eyes, they detect carcasses from a great distance. However, they also eat fish, small mammals and snakes.

For reproduction, they usually lay only one egg in hollow tree stumps or on a branch fork. Both adults care for their offspring for up to 3 months.

8.5.11 Hoatzins *(Opisthocomus hoatzin)*: (Opisthocomidae)

These remarkable birds can often be seen in the early morning and evening hours sitting in trees near the shore. They constantly communicate with each other with a peculiar buzzing and hissing sound, as if an old steam locomotive is blowing out steam. Thus their Aztec (and English) phonetic name is derived from these vocalizations: In German they are also called—today no longer politically correct—gypsy birds because of their maroon, 8-cm feather crown on the head or also skunk birds, since fermenting gases escape from their forestomachs. To the famous German biologist Berhard Grzimek, this pervasive smell reminded him of cow barn scent. The actual stomach is 50 times smaller than the crop. These birds grow to about 70 cm. Their small head with a long neck, long, broadly fanned tail and oversized wing wings are striking. These are black and reddish-brown in colour with white stripes down the neck. The neck and breast are creamy light brown in colour. The red iris is surrounded by bare blue skin (Fig. 8.57).

Fig. 8.57 The hoatzins communicate with each other with a hoarse buzzing sound

Hoatzins fly awkwardly and ponderously. A series of powerful wing beats is always followed by a gliding flight, which ends after a short distance in the branches. Often they just hop from branch to branch in dense foliage, often causing feather damage. The reason for their handicap lies in their pectoral muscles, which are greatly reduced in favour of a greatly enlarged, muscular crop, which serves as a forestomach and is used to ferment their food. The filled crop can account for up to 25% of their body weight. Their favourite food is the large, coarse leaves of the tropical marsh calla (*Montrichardia arborescens*; Sect. 6.3.2), which are also toxic. As these are also low in nutrients, hoatzins have to eat a lot of them, their bacterial flora helps with fermentation and the decomposition and removal of toxins, but the whole digestion process takes up to 48 h—highly unusual for birds.

They live gregariously in groups of up to 20 individuals exclusively on the banks of white-water rivers and lakes, where there are extensive populations of the tropical marsh calla (Fig. 6.26). Their hoarse buzzing and hissing, which serves to "sense the mood", can be heard from a distance.

Breeding takes place during the rainy season. The adults, together with their four young birds of the previous year, always build the nest at high tide. Guarding the nest and feeding the new generation is done by the whole family. The nest, built of twigs and leaves, is always located directly on the riverbank in the branches, about 5 m above the water. In case of danger, the young birds jump out of the nest directly into the water, where they first dive away and hide. As soon as they feel safe, they climb back up the tree to the nest with the help of small rounded, movable claws, which are located at the ends of the hand wings. After about 3 months, these claws disappear. Because of the nutrient-poor diet, the nestlings develop very slowly. Only after 50–70 days they are independent.

The taxonomic position of this family has been greatly debated and is still far from clear. Only one fact seems certain: relatives of the Hoatzins lived in today's Brazil as well as in today's Namibia already about 20 million years ago.

8.5.12 Horned Screamer *(Anhima cornuta)*: (Anhimidae)

These impressive birds live as sedentary birds in the expanses of the floodplains, where you can occasionally observe loose flocks of about six individuals from the canoe. They are, however, very shy. They are massive, goose-like in shape, about 80 cm tall, with powerful wings and feet. Their plumage is black with some white speckles, and their lower neck is patterned black and white. Their irises are amber. A thin, elastic, whitish horn-like structure up to

Fig. 8.58 In the horned screamers a thin, whitish, horn-like structure grows on the forehead and on the wings there are two pointed spines

15 cm long, projecting vertically from the forehead, is conspicuous (Fig. 8.58). This "horn" continues to grow continuously, but the tip often breaks off. Horned screamers also have two pointed spurs on each wing, which they use in territorial fights and for defence.

Their loud calls have earned them the English name "screamers". They feed exclusively on water plants. For reproduction, they build a nest on the ground, in which up to six eggs are laid.

8.5.13 Hummingbirds: Hummingbird Family (Trochilidae)

Hummingbirds exist with about 360 species only in the two Americas from Alaska to Tierra del Fuego. In the Amazon lowlands more than 40 species live. Their plumage colouring is always bright and iridescent. This metallic sheen makes them invisible to predators in the constant play of light and shade in the trees, and their bodies seem to dissolve *(somatolysis)*, which can be seen

Fig. 8.59 Hummingbirds, such as the plain-bellied emerald must visit up to 3000 flowers a day to avoid starvation

well in the **plain-bellied emerald** *(Chrysuronia leucogaster)* (Fig. 8.59). Their beaks, which are always long, vary greatly in size and shape so that they are adapted to the particular food flowers. The extremely long tongue is extended far forward so that its split and hollow tip reaches the nectar at the base of the flower, which is sucked out. The birds have a high demand for energy-rich food, and the sugary nectar fits the bill. They spend the night at only 18 °C body temperature and a 90% reduction in body functions in a kind of suspended animation to avoid starvation. The first rays of sunlight of the beginning day wake them up, and they begin their search for food fully ready for action, having to seek out up to 3000 flowers per day. To meet their protein needs, they also eat spiders and insects, which they steal from spider webs. Every day they have to eat 50% of their body weight in food. This is because their basal metabolic rate is extremely high:

- When shaking in front of a flower, they make up to 50 wing movements per second;
- with the short courtship flights there are even 200 wing movements per second;they have the largest brain of all animals relative to their body weight, namely 4.2%;

- their body consists of a third of flight muscles;
- they possess a lung three times as large and a heart three times as large as a domestic pigeon, relative to body weight;
- their heart beats 500 times per minute (in humans 60–90 times);
- they can fly more than 100 km per hour for a short time and
- they are the only birds that can fly backwards and on the spot.

Their reproduction is very special for birds: they practice a "no-marriage"—there is no pair-bonding during copulation and during the building and rearing of the young. The female builds the nest alone in a bush, a small bowl open at the top made of plant fibers held together with spider threads. After mating, male and female remain strangers to each other, except for brief swoops by the male in front of the sitting female. The clutch consists of two eggs, the female incubates them alone and raises the young birds, which have to be fed up to 140 times a day.

8.5.14 The Anis: Cuckoo Family (Cuculidae)

The distinctive features of these birds, common in the Amazon lowlands, are: uniform black plumage, powerful bills, long tails. The anis vagabond during the day in open habitats such as floodplains, open riverbanks, clearing areas in search of termites, insects, fruits and seeds. They always travel in groups of about ten individuals. Their typical flight pattern consists of alternating fast wing beats and short gliding phases. These birds are very social and look for food, care for their feathers, defend themselves against enemies and care for their broods in groups. They are not brood parasites, but they do engage in a preliminary form of this behaviour: All females lay their eggs in a communal nest, but the highest-ranking individuals sometimes rearrange the eggs in unobserved moments so that their own eggs are on top, or they even throw eggs of other females out of the nest.

The **greater ani** (*Crotophaga major*) grows to the size of a magpie (45 cm) and has a prominent hump on its upper bill. Its smooth plumage shimmers metallic blue (Fig. 8.60). The **smooth-billed ani** (*Crotophaga ani*) is one-third smaller at 35 cm, its powerful upper bill raised like a comb. On the head, neck and throat, the feathers look "shaggy" and the metallic-blue sheen is not present (Fig. 8.61).

Fig. 8.60 The greater ani has a prominent hump on the upper beak

Fig. 8.61 In the smooth-billed ani the upper beak is raised like a comb

8.5.15 Icterids: New World Black Birds (Icteridae)

This family of birds, rich in genera and species, includes two similar species, which are very common on the banks of the rivers and lakes of the Amazon lowlands, but also like to establish their breeding colonies in large trees in the villages and small towns near the rivers. One is the **yellow-rumped caciques**, *(Cacicus cela)*. These birds are not choosy in their choice of tree for a future breeding colony—this can be a small tree standing in the *várzeas*, which is submerged almost to its crown in water during the rainy season. However, one thing seems to be important for colony establishment: The proximity to a wasp nest offers them protection from, for example, predatory toucans and black caracaras, which like to raid their nests.

Their nests are pouch-shaped, up to 40 cm long and flared at the base. They are woven from blades of grass, whereby the nesting material protrudes crosswise from the nest body. Such a colony grows over the years and may well consist of 60 or more nests (Fig. 8.62a). The nests of the previous year are reused and—if necessary—repaired by the females. The clutch consists of only one egg. There is lively traffic in such a colony. The nests are being built, and birds are constantly flying off to search for food or returning from it. They eat insects and fruits. While doing so, the males constantly call with a wide

Fig. 8.62 (**a**) Breeding colony of yellow-rumped caciques, (**b**) Breeding colony of crested oropendola

Fig. 8.63 Yellow-rumped caciques are black in colour except for yellow patches on the wings, rump and tail

variety of sounds such as flutes, cackles and whistles, but they also imitate a wide variety of sounds. If the observer keeps quiet, he can observe this activity extensively and take good photos.

These lively birds are about 30 cm tall, completely black in colour except for a few yellow feathers on the wing covers and rump. Their iris is light blue and their large bill ivory (Fig. 8.63). Their generic name, *Cacicus*, designates a leader or village chief—probably because these birds are so dominant.

The second icterid is the **crested oropendola** *(Psarocolius decumanus)*. It looks very similar to the yellow-rumped cacique but is much larger at just under 50 cm. It is also black, except for the tail feathers. These are maroon at the root of the tail and the butt is sunny yellow. Its iris is pale blue and its powerful bill ivory (Fig. 8.64). The male wears a barely discernible hair-like cap (hence the English name *Crested Oropendola*).

The loud calling of both sexes sounds like a loud clacking. They are colony breeders, and only on very high trees. There they build enormous, elaborately woven hanging nests up to 1.2 m long (Fig. 8.62b). These are tubular until they expand spherically at the base. A dominant male of the colony mates with most of the females after a previous elaborate ritual with the other males of the colony. The clutch consists of two eggs. This bird species also lives in rainforests at the edge of rivers and lakes. It avoids *terra firme* forests.

Fig. 8.64 The crested oropendolas are black in colour except for the maroon root of the tail and the yellow butt

Tip

To photograph birds, you need a powerful telephoto lens with a focal length of 400 or even better 600 mm. There is an app for mobile phones called "Birds of Brazil" that you can download for free. You can use this to identify the birds and even attract them. The app even includes the songs and calls of the birds described. By playing these, you can attract the males of territorial bird species such as toucans and take good photos of them.

8.5.16 New World Flycatcher: Tyrant Flycatcher Family (Tyrannidae)

This large bird family of the two Americas includes more than 430 species. Exemplary three of them are presented here, which the short-time visitor of the Amazon lowland can observe from the boat or also on land. By the way, they owe their name to their aggressiveness towards intruders in their own territory, even towards significantly larger birds of prey. From their perch, they hunt mainly insects.

The **great kiskadee** (*Pitangus sulphuratus*) is one of the most common birds in Amazonia, one can almost call it a "follower of civilization" because it is at

Fig. 8.65 Great kiskadees are called *Bem-ti-vi* (= "I saw you well") in Brazilian

home in the villages and small towns along the Amazon as well as in habitats near the river bank. Its unmistakable call can be heard everywhere. Brazilian speakers call it *Bem-ti-vi* (in English = "I saw you well"), French speakers call it *Qu'est-ce qu'il dit* (in German = "What did he say?"). Among Spanish speakers, it is called *Kiskadee* ("As it calls").

This 22-cm sparrow has a black and white streaked head, medium brown wing coverts, white throat and sulphur yellow belly. Its black insect bill is short and thick (Fig. 8.65). In addition to insects, it catches small fish by diving as well as frogs, lizards, and small crabs, and it also eats fruit. It also breeds in human settlements, building its nest from plant stems, the female incubating the two to three eggs.

The **fork-tailed flycatcher** (*Tyrannus savana*) has a body length of just under 10 cm, but its distinctive forked tail is three times as long in the male (Fig. 8.66). Thus it has the longest tail of any bird on our planet relative to body size. In the female, the tail feathers are much shorter. Its underside is white. Its upper side is grey, the wing coverts medium brown. There is a black cap on the head. It prefers open habitats such as gallery forests, scrublands, grasslands and *várzeas.* The female builds the cup-shaped nest in bushes, the clutch consists of two to three eggs.

The **white-headed marsh tyrant** (*Arundinicola leucocephalis*) grows to 13 cm, the male is blackish-brown in colour with a white head, the underside of the bill is yellow (Fig. 8.67). The female is not as contrastingly coloured: The wing coverts are brown, the head and front dull white, and the tail black. It prefers to live in floodplain meadows, where it hunts insects from its perch on a bush. It builds its grass nest on the end of a branch above water. The two to three eggs are cared for by both parents.

Fig. 8.66 The tail of the male fork-tailed flycatcher is three times as long as its body

Fig. 8.67 **The** male of the marsh tyrant has a white head with otherwise black plumage

8.5.17 Common Potoo *(Nyctibius griseus)*: Potoos (Nyctibiidae)

A night excursion is one of the highlights of any trip to the Amazon. When scanning the river banks with a good spotlight, you can observe numerous birds. On the one hand, there are the diurnal birds that spend the night on trees close to the river bank, for example, kingfishers, great egrets, snowy egrets, capped herons, boat-billed herons and anhingas (darters), which get into the spotlight and do not fly away, as they cannot see anything at night. So you can get very close to these birds. On the other hand, you can observe nocturnal birds—such as the potoos and nightjars. If the potoo gets caught in the headlights, its two eyes reflect red like glowing coals. Like the spectacled caiman, it has a reflective skin layer behind the retina (Sect. 8.4.3). Because this bird is aware of its extreme camouflage, it can be approached quite closely by boat before it skims or simply closes its eyes. During the day, it rests on the top of broken slender tree trunks with its head pointing upward like a stake and its eyes nearly closed (Fig. 8.68). Since his large eyes have a strikingly

Fig. 8.68 The common potoos are perfectly adapted to their environment—during the day and also at night

bright yellow iris, he keeps his eyes closed so that they will not betray him. Nevertheless, he can still see, for there are two vertical slits in his eyelids. With this posture and its grey-brown striped and spotted plumage, the bird blends in with the bark structure of the tree. The camouflage is really perfect! With the good eyes of the boatmen and a laser pointer, I was able to see and photograph this interesting bird during the day. At night it is just as perfectly camouflaged as during the day, even at night it closes its eyes more often.

Its body is strongly built at a size of 40 cm, the tail is long. There are some bristles at the base of the beak and it opens its disproportionately large maw when defending itself against enemies. It lives in open forest habitats at the edge of calm waters. There it goes hunting from its perch at the onset of night, swooping down on large insects such as moths.

To reproduce, an egg is laid in the cavity of a tree stump, both adults take turns brooding, the male broods during the day, the female at night. When canoeing under a full moon and hearing its deep, mournful call *(po-ou po-ou po-ou)*, the experience is certainly a highlight of any trip. The river people also call him ***potoo*** for this reason. English speaking visitors call them also "*poor-me-ones*" after their haunting calls. Another native name is **Urutau** (= "day sleeper"), which translates to "ghostly mouth", because of its large maw. Potoos are related to the swallows and nightjars ("goat-suckers") respectively.

8.5.18 Common Nighthawk *(Chordeiles minor)*: Nightjar Family (Caprimulgidae)

During the night excursions, visitors can observe—especially before the rain or while it is raining—nightjars, of which there are very many different species. In the spotlight they fly with fluttering wing movements just above the water surface. In direct light, their eyes reflect a glowing red. The *common nighthawk* grows to 23 cm, has a dark brown or grey ground colouring with black-grey-brown patterning with some incomplete white bands on the wings. The narrow wings bear a white band. In males, the throat is white. When closed, the bill looks unusually small (Fig. 8.69). During the day they sit crouched lengthwise on a low branch, which camouflages them perfectly. At night they hunt winged ants, moths, moths, fireflies and beetles from their perch. Their irregular zig-zag flight, sometimes interrupted by a dive, is striking. I have often approached their perch by boat with the headlights on. The nighthawk remains sitting on its branch as if paralyzed, you can photograph it at close range and could even grab it—as if it is aware of its perfect camouflage.

Fig. 8.69 The common nighthawk flies over the water surface in an irregular zig-zag flight

8.6 Mammals

Concerning the mammals, as for many other animals, there are indeed many different species, but due to the dense vegetation on the banks of rivers and lakes or in the rainforest itself, they are difficult to spot and rarely observed, although there are over 400 different species of this vertebrate class in Amazonia—in Germany, there are for example only 100.

According to the guiding criteria for this volume, only those animal species are presented here that the visitor can really see during a stay of about 2–3 weeks. This can be in nature itself or also when visiting the *Caboclos* or the indigenous people, because they keep more often animals of the rainforest as "pets". However, they have a different relationship to these animals than we do in our culture. With us, for example, dogs or cats live as part of the family in our household, the animals usually have a name and the owners have a "personal relationship" with them. In the Amazon lowlands, animals living in or around the house take on a different status. They tend to live apart from the family with no name of their own, monkeys and sloths are usually kept on very short leashes. They are usually young animals whose parents have previously ended up in the cooking pot. Highly sensitive animals like monkeys suffer from these experiences and can rarely lead a species-appropriate life. During my visits to such villages over the course of 20 years I have met many different, almost always young mammals from the rainforest as "pets": for example capybaras, (kinkajous), (honey bears), coatis, forest dogs, tapirs, agutis, sloths or also numerous monkey species, among these—besides the species presented here—also woolly spider monkeys (also known as muriquis) as well as different marmosets (zaris), golden lion monkeys, pygmy marmosets and different tamarins.

8.6.1 River Dolphins: Families—River Dolphins (Iniidae) and Oceanic Dolphins (Delphinidae)

Two completely different species of river dolphins live in the Amazon river system as a result of the profound geological events that occurred some 25 million years ago, which separated a huge ocean bay from the Pacific Ocean and gradually transformed it into a freshwater inland lake (Sect. 2.1). These represent relicts of two dolphin families that originally lived in the oceans.

The larger of the two is the **Amazon river dolphin** *(Inia geoffrensis)*, which is also known as the **pink river dolphin** and as the **boto** by river people. It is commonly seen in all types of water, whether white water (like the Amazon), black water (like the *Rio Negro*) or clear water (like the *Rio Tapajós*). During high water, the boto even swims into floodplain forests, where it looks for prey among the branches of the treetops. In doing so, it benefits from the fact that

Tip

There are now some *caboclo* families who regularly feed the botos with fish from their houseboats. This is how they have gotten them used to humans. For a small fee, visitors can swim with the botos in this way and even touch their leathery soft skin (Fig. 8.70). With snorkel goggles, you can also observe them up close underwater.

Fig. 8.70 The pink river dolphin is called "boto" by the river people

its neck and cervical vertebrae are not fused together, so that it can turn its head—like we humans do—and thus avoid obstacles better. Botos are quite curious and often appear right next to the canoe, nevertheless, they are difficult to photograph, most likely to film.

The boto reaches a body length of just under 3 m with a weight of 200 kg, whereby the males are larger and heavier than the females. The body colouration is pink. A roundish head with a clearly separated snout sits on its massive body. Its spherically bulging forehead contains the melon, which is responsible for echolocation. Like all mammals, dolphins have lungs. They breathe with the help of a breathing hole that sits transversely on its head behind the melon; every few minutes they come up to the surface of the river to catch their breath. Incidentally, I have very rarely seen air jumps, as oceanic dolphins more commonly do. Its dorsal fin is barely noticeable as a small bump on his back, while his broad pectoral fins are particularly conspicuous and give him excellent control in the water. The caudal fin is also powerful, enabling the boto to make headway even against the strong currents of the Amazon. Its tiny eyes are not very powerful. Its beak-shaped mouth is densely packed with two rows of teeth each in the upper and lower jaws (Fig. 8.70), with different tooth shapes, which is unique for dolphins.

Its prey spectrum includes over 50 different fish species as well as turtles. To track down its prey, it swims in pairs or in small groups for 30–40 km along the rivers every day for 7–10 h, as documented in a doctoral thesis on its way of life. They are most often encountered where the dark blackwater rivers join the whitewater rivers. At these places, fish from up to three degrees Celsius warmer and acidic black water meet the chemically almost pure and almost ph-neutral white water, which initially irritates and stuns them, making them easy prey for the botos. Echolocation allows them to find their prey perfectly even in the nearly opaque white water. To do this, the melon emits sound waves—like an echo sounder—which are reflected by the fish. The sound waves—returning in the melon—are evaluated in terms of the size, shape and consistency of the prey. Their sonar system is so accurate that they can locate something edible the size of a two-cent coin at a distance of ten metres. In addition, they have tactile hairs on their snout for fine orientation.

Botos have a baby only every 2 years, which is initially anthracite in colour and remains with its mother for many months. During courtship, some male river dolphins decorate themselves with sticks (according to Vera da Silva, Instituto Nacional de Pesquisas de Amaẑonia, *Manaus*).

In Colombia, this river dolphin is unfortunately increasingly hunted. Its meat is used as bait to catch certain catfish, which are very popular as food fish. The river people, however, generally leave the botos alone and do not catch them for food. *Caboclos* say of them that it brings misfortune on their family if one accidentally kills a boto, such as while fishing, when they become entangled in the nets and suffocate.

There are also numerous legends surrounding the boto, which is said to have a number of human-like behaviours: For example, the boto, wearing a white suit and a hat to cover his breathing hole, is said to go ashore at full moon to look for young girls. After seducing one, he disappears back into the river. Even today, there are still many birth certificates in the Amazon lowlands in which, according to the girl, "Boto cor de Rosa" (= pink boto) is entered as the father.

Incidentally, the pink river dolphin was discovered by the French naturalist Etienne Geoffrey Saint-Hilaire in 1807, which is why it bears his name in its honour.

The second Amazon dolphin is the anthracite-coloured **bufero gris** *(Sotalia fluviatilis)*, which remains much smaller at 1.5 m body length and 50 kg (Fig. 8.71). It is seen less frequently, but then usually in small "dolphin schools" in which they swim in synchrony and take a breath. Their behaviour is very similar to oceanic dolphins as they belong to this family. They have a distinct, 15 cm high dorsal fin. The proper indigenous name *Tuxuci* comes from the *Tupi language*.

Fig. 8.71 The Tucuxi dolphin (or Sotalia) is similar in appearance to the marine dolphins

8.6.2 The Pale-Throated Sloth *(Bradypus tridactylus)*: Sloths (Bradypodidae)

We humans have given this animal an arrogant and derogatory name. The naturalist Charles Darwin, for instance, wrote, "It is the ugliest creature that creation has ever produced. Of all animals, this pathetic, crippled creature clings most tenaciously to life." This was unfair because sloths are actually perfectly adapted to their environment and do fantastically well in the rainforest. After all, their food doesn't run away from them either. Their slowness, inconspicuousness and camouflage are their best protection! Recent field studies have also determined that sloths sleep far less than previously thought: It's 9.6 h per day, not up to 20 h as previously thought. Sloths are fairly common inhabitants of the trees near rivers.

Tip

Even if a sloth is sighted from the boat, it is very difficult to communicate the exact position of the animal to the fellow passengers. In this situation a laser pointer is helpful, with which you can show the sloth "spot on".

In this habitat, they also find their favourite food: the leaves of the snakewood (Cecropia), (Sect. 6.1.1). Due to the abundant food supply, these animals have become grazers of the treetops, i.e. true "browsing grazer". As they have neither canine nor incisor teeth, they tear off the leaves with their strongly horny lips. Their grinding teeth, which have a lot to do in this way of feeding, are constantly wearing out, but this is compensated for by permanent regrowth. Due to their special anatomy of thoracic vertebrae and ribs, they can turn their neck 270° vertically and even 330° horizontally, which allows them to pluck leaves without elaborate movements. Like cattle, they are ruminants, as the cecropia leaves, in particular, are difficult to digest. After predigestion in three stomachs, the food enters the main stomach for final fermentation. They breathe only eight times per minute and to further conserve energy, the body temperature drops to 24 °C at night. They live solitary preferentially in the crowns of cecropia trees, where they are difficult to spot. If you specifically scan these pioneer trees with binoculars, you will often see them hanging upside down from the branches. Without binoculars, they look like a bunch of dry leaves from a distance.

Fig. 8.72 The fur of sloths is remarkably dense and shaggy

Adult sloths are up to 80 cm tall and weigh about 10 kg. Their dense, shaggy fur (Fig. 8.72) is a basic gray colour, and males have additional yellowish-white stripes and spots on their bellies. The hairs are parted away from the belly so that rainwater can run off there. Nevertheless, their fur gets wet with every rain shower. During a coat examination, 978 beetles from four different species, nine butterflies and moths, six ticks and ten mites were found.

They only go to the forest floor to defecate—but only about every 8 days due to their slow metabolic activity, which is 48% lower than that of other mammals. While they bury their excrement, pyraloid moths respectively snout moths living in their fur manage to lay their eggs there. The larvae of this month feed on the excrement and the adult moths hatching after metamorphosis seek out the fur of a sloth again. There is a real symbiosis between the both species because the sloths also profit from the moths. The metabolic products of the moths feed green algae, which live in the fur of the sloths during the rainy season. The green algae not only serve as additional camouflage for the sloths, but they also provide the sloths with important plant secondary metabolites that they do not obtain from their other food.

The head of the sloth is small and roundish, the small ears are hidden in the fur, their vision and hearing are poor. A short tail, 10 cm long sits on the torso. Due to their constant head-down posture, their liver and stomach are displaced 135° compared to other mammals, thus the liver cannot press on the stomach. The front extremities are conspicuous, being one and a half times as long as the rear ones. The hands and feet have three digits, armed with long, sharp, arched claws, with the middle claw being the largest and most powerful. They are formed in a way that these animals hang from the branch without any active use of energy. I had once held a sloth by its two arms with my hands. In its defense, it bored the claws of its one hand into my forearm. It took some effort on the part of my fellow travellers to free me from this attack.

Sloths live almost their entire lives in trees, even mating and giving birth to offspring take place there. They change trees every few days and can reach a speed of 5 m per minute. They can also swim surprisingly well and thus sometimes cross a river. Sloths give birth to only one cub per year after a gestation period of 6 months, which is just under 20 cm tall at birth.

Unfortunately, the sloths are hunted in the Amazon lowlands, as their meat is popular with the river people. The adult animals are eaten, the young are kept as "pets", but since they are food specialists, it is difficult to keep them. By the way, the closest relatives of sloths are anteaters.

8.6.3 Common Squirrel Monkey *(Saimiri sciureus)*: Capuchin Family (Cebidae)

Among the approximately 60 monkey species of the Amazon basin, the common squirrel monkey is widespread. Recent genetic research methods have led to a splitting of the genus *Saimiri* into more and more species and subspecies. Meanwhile, zoologists already distinguish about up to ten different species. Therefore, the species described here, which is native to the region north and south of the Amazon, is now called the common squirrel monkey.

These monkeys remain rather small with a body length of about 35 cm, in addition, the tail is about 40 cm. The males are slightly larger than the females. Their body ground colouration is greenish-brownish, the belly is lighter coloured, the lower extremities are orange-yellow.

Their food spectrum is broad and ranges from insects and termites to bird eggs, leaves, flowers, buds and nectar. They prefer to stay in flooded forests *(Igapós)* or on trees close to the river bank, where they forage in groups of up to 30 animals in their large territories of up to ten square kilometres. These vast territories ensure their survival in periods of food shortage.

Their social structure is unusual for monkeys: the females form a fixed hierarchy around which the other group members (males, young) rank

Fig. 8.73 The mask-like facial markings with black snout and dark eyes is a distinctive feature of the squirrel monkeys

themselves. During the reproductive season, the dominant males sire most of the young in the group. All females take care of the new arrivals.

The mask-like facial markings with a black snout and dark eyes gave them their German name "skull monkey" (Fig. 8.73). The inhabitants of the Amazon lowlands, however, call these monkeys "Macaco de cheiro", in English "stink-monkey". They regularly let urine run over their hands and then rub their feet and fur with it. With this intensive smell, they lay a "scent road" in the foliage on their way through the treetops. They use it to mark their territory and find their way more quickly if they have to flee quickly. The generic name *Saimiri* is in turn a proper name in the *Tupi language.*

8.6.4 White-Fronted Capuchin *(Cebus albifrons)*: Capuchin Family (Cebidae)

Of the approximately 15 different species of the genus *Cebus* occurring from Central America to Argentina, the one described here is native to the Amazon

Tip

In the Amazon lowlands there are several places where you can observe and photograph squirrel monkeys at close range:

- *Isla de los Micos* (Colombia), a river island in the middle reaches of the Amazon, about 1.5 h (35 km) upriver from *Leticia* by speedboat. There, a large group of about 100 squirrel monkeys are kept in the rainforest. When visitors take bananas from the containers provided, the monkeys come down from the trees without shyness to take the fruit from their hands.
- *Isla de los Monos* (Peru), also a river island in the upper reaches of the Amazon, about an hour by speedboat upriver from *Iquitos*. There is a private rehabilitation centre where monkeys from different species are cared for, for example, howler monkeys, woolly spider monkeys, saki monkeys, uakaris and tamarins. You can observe these animals there at close range.
- If you go up the white-water river *Rio Ariau* from the right bank of the *Rio Negro*, about 50 km from *Manaus*, you can meet groups of squirrel monkeys in the early morning on the right bank of the river with a little luck. They are used to being fed by tourists and when they see bananas, they also come onto the canoes without fear to carefully grab the fruits from our hands.

lowlands. The preferred habitats are gallery forests along the banks of rivers and *várzea forests*, low clusters of trees at the edge of floodplain meadows, and secondary forests. These monkeys are about 40 cm tall, with the addition of the tail, also about 40 cm. The males are slightly larger than the females. Their long and thin extremities are striking. Their strong jaws give them a fierce, bulldog-like appearance. With a bite force of 140 kg, they can crack even hard palm nuts. Their basic body colouration is light brown. Their flesh-coloured face is surrounded by a broad band of cream-coloured hair. The chest is also cream-coloured (Fig. 8.74).

White-fronted capuchin roam in groups of 10–15 individuals in search of food. In doing so, they also go more frequently to the ground, where they are easy to observe. Their diet includes insects, spiders, frogs, lizards, but also

Fig. 8.74 The strong jaws of the white-fronted capuchin give it a fierce appearance

leaves, fruits, seeds and even honey and nectar. The favourite flowers are the showy red and yellow flower spikes of the monkey's brush *(Combretum aubletii*; Sect. 6.2.5). I once observed a large male capuchin monkey on cannonball tree (Sect. 7.2.2) carrying a picked fruit, considerably larger than his own head, to a broad branch. There he struck it several times with both hands on the branch in order to break open its shell. After he had succeeded in doing this, he ate the flesh of the fruit.

In fact, this species of monkey even understands cause and effect: they rub their bodies intensively with essential oils of the peacock flower (Fabaceae), for example, because they have realized that they are bitten less by mosquitoes afterwards.

Both sexes develop a hierarchy in their group, which comes into play during mutual grooming and reproduction. The dominant males sire most of the young. As a rule, a female has only one baby per year. The males participate in the rearing of the young.

The name *Capuchin* refers to the light brown clothing and headgear (the hood) of the capuchin monks.

8.6.5 Columbia Red Howler *(Alouatta seniculus)*: (Atelidae)

There are 12 different howler monkey species, of which the Columbia *red howler* lives in the Amazon basin. You can see them now and then from the canoe in their feeding trees on the river bank. There they spend their resting periods. They need breaks to digest their hard food, which consists mainly of leaves. They have a stocky body up to 70 cm in length and an equally long tail, with the males being considerably larger than the females. The body base colouration of their long fur is reddish-brown and their hairless face is black. Their long beard is also typical (Fig. 8.75). The long tail serves as a powerful grasping hand ("fifth hand"), and its posterior part is hairless on the underside.

Fig. 8.75 The piercing, lion-like call of the howler monkeys serves to mark their territory

Their calls give them their name. Often, especially in the early morning, you can hear their piercing call, which is more like a loud roar, similar to that of the lion. British scientists have found that the smaller the testicles of a howler monkey, the deeper its voice. For sufficient food trees, a group of five to ten individuals needs a large territory. They defend this territory with loud roars from both sexes. Usually, the males start and a neighbouring, more distant group answers. This avoids costly territorial fights, which is also urgently necessary with a low-energy diet. To make this enormous volume possible, the hyoid bone and the thyroid cartilage of the larynx have greatly increased in size during their evolution.

In the group, a strict hierarchy prevails among the males, the dominant males mate with the females, who have only one baby per year, around which all group-members take care.

8.7 Butterflies

For every traveller, it is a special experience to see a **peleides blue morpho** *(Morpho peleides)*; family: Nymphalidae flying in the bright sunlight. The opportunity is great during midday when they like to fly along smaller, open water areas such as *igarapés* or small rivers. However, in the heat of the day, they fly very fast in their typical gauzy flight pattern. The sight is particularly contrasting when they fly over blackwater rivers and lakes. The male reaches a wingspan of 15 cm and its upper wing is bright blue, the outer edges black (Fig. 8.76). The female remains somewhat smaller, and her blue is not as vivid. Interestingly, the blue colour impression is not due to colour pigments. Rather, it is an optical trick that causes the bright blue: Through *interference*, the superposition of several light waves, the grooved structure of the chitinous scales is refracted so that only the blue light becomes visible.

There is also a lotus effect on the wings: Fine grooves with a sawtooth profile run along the edges of the wings, allowing raindrops to drip off the top of the profile. When these butterflies perch on a leaf in the shade, they are almost impossible to identify because they are perfectly camouflaged with their underwings now folded up: the dark brown colouration with wavy white stripes blends seamlessly into the chiaroscuro of the foliage. In addition, there are seven yellow mimicry-edged mimicry eyespots on underwings, a warning costume *(mimicry)* that imitates owl eyes (Fig. 8.76).

These emperors, as they are also called, feed on the juice of overripe fruit, their larvae (caterpillars) feeding on leaves of various genera of the legume family (Fabaceae), for example, *Campsiandra, Clitoria, Inga* and *Cassia*

Fig. 8.76 The upper wing surfaces of the peleides blue morpho are bright blue. This colour impression is created by *interference* which refracts the structure of the wing scales so that only the blue light components are visible. When this butterfly is sitting, only the dark colouring of its underwings with the seven eye dummies is visible

(Sect. 6.1). Eggs are attached to the upper surface of leaves, and the hatching caterpillars undergo several moults until they are nearly 8 cm long. They are yellow and red spotted with a hairy head capsule. After 4 months, pupation occurs and then the moth hatches.

Of many butterflies in the Amazon lowlands, visitors only get to see their caterpillars. However, these are often spectacularly coloured and/or of unusual shape. Thus, they often catch the visitor's eye during his trips along the banks or during his hikes through the rainforest. But be careful: touching or picking them up, for example for a better photo position, can be very dangerous: Almost all of these caterpillars have fine hairs or even short thorns that break off when touched, and their barbs leave them in your skin, where they deliver their venom. This happened to me once with the caterpillar of the **bullseye moth** (*Automeris liberia*). After a stinging hair tip penetrated the first limb of the thumb, it was paralyzed for 2 days. These nocturnal moths, which belong to the Saturniidae, have two large eyespots on their orange hindwings and have a wingspan of 10 cm. The ground colouration of the last caterpillar stage

Fig. 8.77 The caterpillar of the bull's eye moth protects itself with innumerable poisonous stinging hairs

is a rich green with a conspicuous cream-dark red longitudinal line below the tracheal openings. An enormous number of toxic stinging hairs, fir-tree-like on a common shaft, protect the caterpillar from predators (Fig. 8.77).

Tip

The images of the imaginal stage (sexually mature butterflies) of the caterpillars described here can be found, for example, on the Internet at Google Images.

Another spectacular caterpillar of the Amazon lowlands is a hairy structure: 11 tower-like, white knobs sit on the approximately 8-cm-long, light-brown body, on each of which a tuft of fine, dark-brown, several-centimetres-long, poisonous hairs sprout from a red base above and below (Fig. 8.78). These break off easily when touched and cause long-lasting inflammation, allergic reactions, and even paralysis in the skin. The butterfly that emerges from this caterpillar is the **flannel moth** (*Megalogyne lanata*; family Megalopygidae). This is a white-gray, woolly, nocturnal moth with a wingspan of 5 cm. These caterpillars have different food plants, such as the leaves of the cashew tree.

Fig. 8.78 The caterpillar of the flannel moth is surrounded by a tangle of long, poisonous hairs

The caterpillar of the **wool moth** (*Podalia orsilochus*) from the same family (Megalopygidae) also has a sensational shape. The entire body, about 6 cm long, is completely covered with a stiff, dark beige to golden brown wavy fur (Fig. 8.79), which covers the head with its mouth opening as well as the seven pairs of feet, the number of which is highly unusual for caterpillars. These hair-like structures are bristle hairs (setae) with a split-tip. When touched, they release a neurotoxin that causes long-lasting, intense, dull pain. They trigger local toxic effects like inflammation, hemorrhage, necrosis. Sufferers describe them as if a bone has been broken. The associated moth has hairy legs and is covered with a whitish-grey spotted fur with a wing span of 6 cm. Its distribution ranges from the Amazon lowlands to Central America and Florida. The *caterpillar* has recently been nicknamed *Trump Caterpillar* in reference to the hairstyle of the former United States President Donald Trump. The family name *Megalopygidae* is composed of *megalo* (= large) and *pygidium* (= hind end) and refers to the caterpillar's peculiar shape.

The caterpillars of the **tetrio sphynx** also known as "frangipani hornworm" (*Pseudosphynx tetrio*; family Sphyngidae) wear a spectacular warning costume: the head is bright red-orange, the body bears a row of bright yellow stripes on a black background (Fig. 8.80). The anal horn, typical of sphinx or "hawk" moths, is thread-like and swings vigorously back and forth in response to danger.

Fig. 8.79 The bristly hairs of the caterpillar of the wool moth release a nerve toxin when touched. The head is on the left side of the photo

Fig. 8.80 The caterpillars of the tetrio phynx wear a spectacular warning costume

The up to 100 eggs are laid on a food plant, on which the hatched small caterpillars immediately start to eat the leaves. They are extreme food specialists and eat the leaves of only a few plant species from the family *Apocynaceae*, especially the red frangipani tree (*Plumeria rubra*), but also the golden trumpet (*Allamanda cathartica*; Sect. 6.2.3). In order not to die from the poisonous latex of the leaves themselves, the caterpillars first gnaw a hole in the stem so that most of the poisonous latex drains off. Only then do they eat the leaf. After only 30 days, the sixth and final caterpillar stage is complete. The caterpillar now has a body length of 8 cm. After a resting period of a few days, during which pupation takes place, the sexually mature moths hatch out. The large, nocturnal-moth with a wingspan of 17 cm—has a brown patterned body with scaly frosting. The abdomen appears swollen.

The conspicuous signal colouration and the ingested poison protect the caterpillars from predators. But the poison ingested with food must be neutralized by the body with a great expenditure of energy. Therefore, these caterpillars specialize in only one venom. With a wider range of food plants, the energy expenditure would be many times greater, because then different poisons would have to be neutralized.

9

Animal Observations in the Rainforest

Animal observations in the rainforest are even more difficult than on the river banks. As a human, you are surrounded by tall trees both in the *igapós* (flood forests) and in the *terra-firme.* The undergrowth is mostly low, but nevertheless you can only see a few meters ahead without any visual obstruction. Birds of the rainforest are therefore more likely to be seen flying over a body of water (toucans or parrots, for example) or perched on trees close to the river bank. Tree-dwelling mammals are easier to spot in the thinner floodplain forests. In the following, two interesting representatives of poison dart frogs, two reptile species (basilisks and anolis), various bird and monkey species, three representatives of the porcupine family, the tarantula, three important ant species of the rainforest (leafcutter, Aztec, and bull ants), termites, the actaeon beetle and the praying mantis are presented.

In general, there are only a few individuals of one animal species in all these Amazon forests because of the scarcity of food. For every 1000 t of plant biomass, there are only 150 t of animals, 50% of which are ants and termites. The ratio of live weight between animals and plants is also extremely low at 1:6000. This testifies to the scarcity of animals in the rainforest, since the plant mass is 6000 times that of the animals. As a result of this scarce food supply, the basal metabolic rate of almost all Amazonian mammals—even of apes—is only 70–80% of what would actually be required given their body size. A reduction of the energetically very costly basal metabolic rate for the maintenance of vital bodily functions such as body temperature, respiration, heartbeat and circulation by so many percentage points corresponds to energy savings such as we find in modern heating systems.

L. Staeck, *Fascination Amazon River*, https://doi.org/10.1007/978-3-662-64452-2_9

9.1 Poison Dart Frogs

The rainforest, with its high humidity and always constant temperatures, creates ideal living conditions for frogs, which have an unprotected body covered only with mucous membrane. In this twilight of the jungle floor, there are a large number of frogs and toads, most of which have dark body colours and therefore elude the hiker's gaze. One group of frogs, however, is particularly colourful: the poison arrow frogs. There are about 170 species of these, but only a few of them live in the Amazon basin, the majority of them being endemic to the Andes, the regions of the Amazon tributaries Rio Maranon, and Ucayali, and Central America. Most species are very small, barely reaching the size of two two-cents coins. In the western Amazon basin, if you are lucky, you can spot this small poison dart frog on the jungle floor in the region around *Nauta* (west of *Iquitos*). While the female grows to almost 20 mm, the males remain considerably smaller at 12 mm. Its blue-green colour patterns on the abdomen and extremities are striking, as are the alternating black and yellow or yellow-orange stripes on the rest of its body (Fig. 9.1). They live mostly in the foliage layer and climb up into the treetops only during the breeding season. Their sucking discs on their toes make it easier for them to climb. They usually live in small groups of up to six individuals, but at the end of the rainy season many of these small groups join together to form large reproductive communities on the forest floor. The males compete with each other for the females. After laying a maximum of

Fig. 9.1 The belly-spotted poison dart frog signals its venomousness with its striking colour pattern

8–10 eggs on a withered leaf, both parents guard the clutch, which is regularly moistened. The hatched tadpoles attach themselves with the tough mucus of their skin to the back of the female, which carries them in turn "piggyback" up a tree. When female has found a pool of water in a bromeliad, the tadpoles are deposited there and further cared for. They are even fed with unfertilized eggs until metamorphosis into the little frog is complete in 8 weeks. They are named poison dart frog because their skin, in fact, is poisonous. However, they do not make the poison themselves, but they absorb it with their food—namely ants, mites and beetles—and store it in special skin cells under their epidermis. This poison makes them inedible, as their signal colours strongly indicate. Humans can also be dangerously poisoned through open wounds, possibly even through skin pores, which can be fatal. The residents of these areas take advantage of these characteristics of the frogs: the frogs are heated over a fireplace on a grate. Before the frogs die, their skin secretes a whitish foam that contains the poison in a highly concentrated form. Now arrowheads are soaked with this foamy slime and then used to hunt monkeys, sloths, birds and so on.

In the US and Europe there are frog lovers who keep and breed these frogs in terrariums. Wild-caught bring the poison with their food, but after about a year of captivity, these frogs are poison-free, because they are fed with fruit flies.

Another brightly coloured poison arrow frog occurs in the same region: the **red-backed poison dart frog** *(Ranitomeya reticulata)*. This is similar in size to the belly-spotted dart poison frog (Fig. 9.2).

9.2 Lizards

The **striped basilisk** (Basiliscus vittatus) belongs to the basilisk lizard family. Its generic name *Basiliscus* is Greek and means "little king". In mythology, the *basilisk* is a cock-like hybrid creature that wears a crown on its head and whose abdomen is a snake. The species name *vittatus*, on the other hand, translates as "striped." Both are an apt description of the animal:

> the slender lizard has a body length of just under 20 cm and a tail that is significantly longer than the body. It is long-legged, the fingers are strikingly long and thin. Only the males possess a helmet-like protuberance on the head. The body is dark brown in colour with a narrow, cream-coloured stripe on each side starting from the eyes and extending to the end of the trunk, and two equally broad stripes slightly lower, starting behind the mouth and also extending to the end of the trunk Fig. 9.3a).

Fig. 9.2 The red-backed poison dart frog, only 12 mm in size, is also brightly coloured

In the dry season it inhabits the ground and low vegetation of floodplain forests, during the rainy season it lives on trees. It eats insects, frogs, flowers and fruits. The lizard lays its eggs in a hollow in the ground, where they are left to fend for themselves.

The second English name of this lizard is *Jesus Christ lizard.* This is because they are able to run about 4 m across the water when in danger, rearing up on their hind legs. They have skin folds between their toes, which increases the surface area of the foot. Air pockets also formed under the foot, which provides buoyancy. After the first few metres they sink back into the water and have to swim on. Nevertheless, during such an escape they can reach a speed of 1.5 m/s for a short time. I myself have been able to observe this escape behaviour a few times when these lizards were running across a shallow marshy area in the forest.

The **Amazon anolis** (*Anolis punctatus*) belongs to the fringed finger lizards. It is up to 15 cm tall, slender, with a tail one and a half times longer than its body. Depending on the substrate and its mood, this lizard is of green or brown basic colouration with small, barely recognizable white spots. For this reason it is also called *spotted anolis* (Fig. 9.3b). Both sexes have an orange wattle, which they can spread. They can change colour depending on mood

Fig. 9.3 (**a**) The striped basilisk can run up to four metres across the water in case of danger, (**b**) The Amazonian anolis can change its basic body colouring like a chameleon

or surrounding colour like a chameleon. The oval, naked eardrum sits behind the eye. A typical feature is their pointed elongated skull with high-set eyes. They inhabit shady woody stumps and the undergrowth in *terra firme*, where they hunt for insects. They behave territorially, meaning they defend their territory from intruders. Under their feet are adhesive soles that allow them to cling to vertical surfaces.

9.3 Parrots

Visitors to the Amazon lowlands can observe parrots especially when they fly over them—usually in groups—screeching loudly. Thus, the birds keep loud contact with each other even when flying. They prefer to spend the day in clearings and riverbanks of the floodplain forests *(igapós)* and in the *terra firme*. Every morning they take off from their roosting trees to their feeding trees, which are located within a radius of about 10 km.

Most conspicuous are the largest among them, the exceedingly colourful macaws. In the Amazon lowlands the **scarlet macaw** *(Ara macao)* is common. It grows up to 90 cm. Its basic colouring is light red, its wing coverts blue-yellow-red. Its powerful, strongly curved, cream-coloured upper bill with a pointed hook at the end is also remarkable (Fig. 9.4a). This bill serves as a climbing aid, a "third foot" so to speak. This allows this large bird to shimmy from branch to branch, even in dense branches, to reach fruits, seeds, and buds. Its favourite food is the fruit of the *tucumã palm (Astrocaryum),* (see Sect. 7.3.2). From time to time it also eats clay, which it uses to neutralize poisonous plant parts. With its muscular tongue it first examines every food offer before it eats it.

Macaws are very sociable birds, so they are usually in groups of 5–10. In such a group they live in pairs in a lifelong monogamy. As cave breeders they like to look for broken palm trunks, into which they hack a brood cavity with their strong beaks. Of the two to four eggs, usually only one young bird will remain. The female broods, the male provides the food. The young bird remains in the nest for 4 months.

The second large parrot in Amazonia is the **blue-and-yellow macaw** *(Ara ararauna)*, which resembles the scarlet macaw in shape and habits. Its upper-parts are blue, its breast and belly are golden yellow, and its bill is black (Fig. 9.4b).

The second group within the true parrots (*Psittacidae*) includes the Amazon parrots, of which there are about 30 species in the Amazon lowlands and adjacent regions. The medium-sized birds of about 30 cm body length are mostly green in colour, but their distinguishing marks are single yellow, red or blue spots of colour—"badges"—especially on the forehead, on the tail or under the wings, for example in the case of the **blue-fronted Amazon** *(Amazona aestiva)*. The colouration of the head and breast is very variable and may include reddish feathers (Fig. 9.5).

The broad rounded tail and the broad also rounded wings are striking. Their outer tail feathers are red at the base. Due to the opposite position of the toes, their foot can be used as a grasping pincer and is often used to hold their food, for example nuts.

Amazon parrots are also very social and live in larger groups, with adult birds forming a long-term pair bond. They live in rainforests and open cultivated land, where they eat seeds, fruits, buds and flowers. As visitors, we

Fig. 9.4 (a) At 90 cm, the scarlet macaw is the largest parrot in the Amazon. (b) The blue-and-yellow macaw also lives in the Amazon lowlands

Fig. 9.5 The blue-fronted Amazon parrot belongs to the Amazon parrots. The tui parakeet (smaller picture) belongs to the true parrots

usually only see them flying over our heads calling loudly or crossing a river. If we are lucky, we can also observe them at the tops of their feeding trees.

At sunset, the birds fly to their roosting trees, which are usually located near the river bank. Once our ship approached such a roosting tree. From several hundred yards away we could hear their raucous calling, and when we finally reached the tree at the last light of day the noise swelled to a deafening level. There must have been many hundreds of Amazon parrots, which exchanged like in a huge *chat room* at the end of the day about their food trees.

The third group of the true parrots includes the smallest, usually only 10–15 cm large "sparrow of the tropics" the **tui parakeet** *(Brotogeris sanctithomae)*. It is a short-tailed bird about 15 cm in size. Its plumage is rich green on the head, back and wings and light green on the belly. The forehead above the strong, reddish-brown bill is sunny yellow. Its iris is white, the eye ring bluish-white (Fig. 9.5).

These little parakeets are often seen flying noisily over the rivers in large flocks in all types of forests or foraging in their food trees for seeds, fruits, flowers and buds. They are also mated for life. In downtown *Leticia*, I have often witnessed a special spectacle just before sunset: thousands of these small parakeets fly in from all directions to stay overnight on their roost tree, a large fig. Even into the darkness, the noise of the chirping and screeching mini parakeets can be heard far into the side streets.

I have frequently met all three groups of parrots as pets in the villages of the *caboclos* and Indians. They can be easily tamed as young birds and are also quite talented in speech. Thus they whistle melodies or even recite whole sentences.

9.4 Toucans

One of the special experiences of a stay in the Amazon basin is to observe a toucan from the boat. Once a male was so intensely busy with his loud mating calls that we could approach him up to a few meters.

The largest toucan of the Amazon lowlands is the **giant toucan** or *toco toucan*, which grows to 60 cm, is short-winged and long-tailed. Its plumage is glossy black, its neck white. The skin around the eye is yellow. Its huge, red-orange, 20 cm long bill with serrated edges is striking (Fig. 9.6). This is well supplied with blood and regulates the bird's heat balance. It is also excellent for picking or even squeezing fruit and then drinking the juice. The bird can handle the tips of the beak like tweezers.

Its relationship to woodpeckers is evident from its toes, two of which are pointed forward and two backward. It is a pure tree dweller, preferring semi-open terrain.

Fig. 9.6 The giant toucan (*toco*) has a 20 cm long, red-orange coloured beak

Its diet consists of fruits, insects, lizards, bird eggs and young birds. From the water, it is seen mainly when crossing rivers. It travels either singly or in small groups. Its call is unmistakable and consists of a loud succession of rapid tones. To reproduce, it hacks a burrow in dead tree trunks, especially in palm trees. The clutch consists of two to four eggs, and both adults care for the brood. Among the 45 toucan species, more than five live in the Amazon lowlands.

9.5 Woodpeckers

Woodpeckers can be observed quite often from the boat on trees, whereby the *Campephilus*, a genus of large American woodpeckers are native to South America with 11 species. They are particularly conspicuous. The **powerful woodpecke**r *(Campephilus pollens)*—described here—belongs to this group. With its 38 cm body length it belongs to the group of large woodpeckers. The species name *pollens* means "mighty" and refers to this body size.

Only the male has the bright red forehead and head cap, the female lacks the red completely. A white, v-shaped stripe pattern runs from the base of the bill to the back (Fig. 9.7). The breast and belly are banded with blackish cinnamon. The toes are arranged in pairs, with two pointing forward and two backward. The long bill is chisel-shaped. Its large rump, the rear upper part of the back or rather its feathering, serves as a support when it moves along the tree trunk.

The powerful woodpecker lives in all types of forest, preferring riverine and semi-open habitats. To breed, it hacks its breeding cavities into mostly dead tree trunks. The clutch consists of two to four eggs, and both adults care for the brood.

The **cream-coloured woodpecker** *(Celeus flavus)* can often be seen plundering termite nests. Termites are its favourite food, along with tree-dwelling ants. From time to time it also eats seeds and fruits. Its plumage and bill are pale yellow. The wings are dark brown, and the tail feathers black. Its head feathers can be raised to form a crown. Only the male has the large bright red beard patch (Fig. 9.8).

9.6 Tanagers

More than 380 tanager species belong to the family *Thraupidae*, which is found primarily in South and Middle America and the Caribbean. In the Amazon basin there are numerous of these usually very strikingly coloured

Fig. 9.7 With its 38 cm, the powerful woodpecker is one of the large woodpeckers

Fig. 9.8 In the cream-coloured woodpecker, only the male bears the red beard spot

birds. They live on moist forest edges, but are difficult to observe in the foliage of the trees. They eat insects, for example wandering ants, flying termites, and spiders. The name *tanager* is of Indian origin.

The **masked crimson tanagers** *(Ramphocelus nigrogularis)* are quite easy to spot with their colourful plumage and a body length of 17 cm in the foliage. Their head, neck, nape, breast and flank are bright crimson, the rest of the plumage is black, the short, strong bill is composed of a black upper part and a white lower part (Fig. 9.9). These birds live in pairs or in smaller groups. The cup-shaped nest, made of dry leaves, is built in tree canopies. The clutch of two eggs is taken care of by the female.

The species name is composed of *nigro* (= black) and *gula* (= throat) and refers to the black throat of the bird (Fig. 9.9).

Tip

If you want to observe or photograph birds, it is best to follow a flying bird with binoculars until it perches on a branch. Then you approach slowly its resting place diagonally from the side—never directly—until you can take a good photo.

Fig. 9.9 The masked crimson tanager has predominantly red plumage

The **blue-gray tanager**, nickname: "blue jean", *(Thraupis episcopus)* is quite common along riverbanks and in secondary forests. The plumage of the 15 cm large bird is slate-grey, the wing coverts green-blue to turquoise-blue. Its nest, built from a variety of plant material, is housed in low trees, the clutch consisting of up to three eggs.

This bird is also called bishop's tanager (the species name *episcopus* means "bishop"): the first description in 1760 was by the French, devout zoologist Jacques Brisson, who called this bird "bishop's bird".

9.7 Monkeys

The **pied tamarins** (*Sanguinus bicolor*) belong to the family *Callitrichidae* including marmosets and tamarins. They have a body size of up to 28 cm with a tail length of 40 cm. Except for the big toes, the fingers and toes are provided with claws. The hairless, black head with large ears is striking. Long, white fur covers shoulders, neck, chest and arms. The rest of the body is light brown, the long tail is black with a light brown tip (Fig. 9.10).

These tamarins inhabit a small rainforest area in the Amazon basin near *Manaus*, the starting point for many Amazon journeys. There they live mostly high up in the treetops in groups of 3–6 members, where they are diurnal and search for flowers, nectar, fruits, tree sap, insects, spiders and birds' eggs. With

Fig. 9.10 In the pied tamarins, the alpha female mates—most unusually—with several males

them, a rare phenomenon can be observed in mammals: In the group consisting of several females, males and young, one female dominates—normally in mammals, one male dominates. Only this alpha female mates with several males of the group. Usually twins are born, which are cared for by the males. Because of its narrow habitat, this species of monkey is critically endangered by deforestation of the rainforest and by the increasing urbanization of its habitat. The term *tamarin* is used in the *Tupi* language to refer to these monkeys.

Tip

When visiting the *Tropical Hotel* on the western outskirts of *Manaus (Avenida Coronel Teixera)*—very close to *Praia da Ponta Negra* on the banks of the *Rio Negro*—after breakfast you can go into the immediately adjacent rainforest to observe these interesting tamarins. There are some paths through this small forest from which one can scan the trees with the binoculars. Especially in the morning hours I was able to observe and photograph this endangered monkey species. Presently the Tropical Hotel is closed. Hopefully it will reopen soon—probably under a different name.

9.7.1 Spix's Night Monkey *(Aotus vociferans)*: Night Monkey (*Aotidae*)

These only nocturnal monkeys of South America live in the depths of the rainforest on trees, where they sleep during the day in tree hollows. Large rivers form insurmountable distribution barriers for them. They are about the size of a squirrel (35 cm) and their tail measures 40 cm. Their soft, dense fur is mottled gray to brown-red, and their chest is yellow-brown. Three black longitudinal stripes run across the face to behind the roundish head. The large eye sockets with the large brown eyes and the long fingers are striking (Fig. 9.11).

Spix's night monkeys live in a single-family group, which means that both parents inhabit a tree hollow with their offspring until the offspring has become independent. They live strictly territorial and communicate with each other with owl-like calls. Shortly after sunset, they leave their roosting cavity to feed. They are only capable of black and white vision, but their sense of hearing and touch are well developed. They hunt tree frogs, night insects,

Fig. 9.11 The large eyes of the Spix's night monkey reveal that this species is nocturnal

spiders, also eat bird eggs as well as leaves, flowers and fruits. The female gives birth to one young per year, which is cared for by the male.

The genus name *Aotus* is derived from the Greek *aoteín* (="sleeping deeply"). The scientific species name *vociferans* translates as "calling", its name goes back to its discoverer, the German explorer *Johann Ritter von Spix*, who took part in an expedition to Brazil between 1817 and 1820 together with the German naturalist *Carl Friedrich Philipp von Martius*. Due to his extensive collections and publications, *Johann Ritter von Spix* is also called the "Bavarian Humboldt".

Tip

On the grounds of a small hotel in *Novo Airão* called Amazonia Park & Suites (its former name was *Pousada Tarântula*) you can observe a family of Spix's night monkeys shortly before sunset. In a jungle-like area just about 100 m from the hotel, a family of this species lives in a tree hollow. When the day ends, they peer out of the entrance of this cave (Fig. 9.11). *Novo Airão* is located on the *Rio Negro*, about 195 km (3 h by car) from *Manaus*.

9.8 Porcupine Relatives

If you walk attentively through the rainforest *(igapós* and *terra-firme)*, often looking up into the treetops, you can observe representatives of the porcupine relatives on the branches. There are dozens of different species of New world porcupines, tree rats and spiny rats, which are difficult to identify if you only see them briefly. Generally, they have stocky bodies and dense, long, brown fur. They spend their entire lives in trees, so even prolonged floods don't bother them. At night, they sleep in hollow logs or tree cavities. At dusk they come out to eat fruits, leaves, saplings, bark, roots and tubers as well as tree-living small vertebrates. So far, there are no reliable findings about their reproduction.

The **Brazilian porcupine** *(Coendou prehensilis)* belongs to the family of the tree porcupines *(Erethizontidae)*. It looks quite massive and grows up to 60 cm. In addition, its tail is about 45 cm long. It is also called *coendu*. Its entire body is covered with black and white coloured thorn-like spines up to 5 cm long, and its tail is spikeless, but modified for grasping (Fig. 9.12). Its fore and hind feet are also adapted for grasping. Thus it is amazingly quick on

Fig. 9.12 The Brazilian porcupine is an arboreal inhabitant of the *terra firme*. The brush-tailed rat (smaller picture) is not related to the rats, but belongs to the New world porcupines, which form a separate family

the run—I was able to see this for myself when we surprised it on a low branch. It is a loner and active at twilight.

The **yellow-crowned brush-tailed rat** *(Isothrix bistriata)* resembles rats externally, but is not related to them. Rather, it belongs to the family of spiny rats (*Echimyidae*) and has bristly hairs on the back and sides. With a body length of 27 cm and a tail just as long, it is equipped with short broad feet that have curved claws. This makes it excellently adapted to life in trees. It prefers trees close to river banks that have caves for roosting. The tail is bushy like that of squirrels. These animals are also crepuscular and solitary (Fig. 9.12). The genus name is composed of *isos* (= equal length) and *thrix* (= hair).

Another porcupine relative is the **golden-rumped agouti** (*Dasyprocta leporina).* It grows to the size of a rabbit, is long-legged, and with its short, rough, almost golden fur and hairless tail (Fig. 9.13) can be described as a streamlined forest hatchling. As a ground-dwelling toe-walker, the aguti can gallop away in case of danger, for which it uses self-made trails in its permanent territory in the dense rainforest.

They are purely vegetarian and eat leaves, stems, roots and fruits. To feed, it sits on its hind legs and takes a fruit in its front paws. It is the only animal in the rainforest that is able to open the extremely hard-shelled Brazil nut with its strong incisors to get at the tasty, protein-rich seeds. They also lay in stores, which they bury, helping to spread the Brazil nut tree (Sect. 5.8). They are solitary and crepuscular. Only for mating do agoutis congregate for short

Fig. 9.13 The golden-rumped aguti is the only animal in the rainforest capable of opening the Brazil nut

periods—the young can walk within an hour of birth. Agouti meat is often offered at the local village markets, which the *caboclos* like as a change from the constant fish meals.

9.9 Tarantulas

The first encounter with a tarantula is for many visitors to the Amazon lowlands a highlight of the trip—or a great stress—if you suffer from *arachnophobia* (fear of spiders). If you know where they hide during the day, you can easily coax them out of their hiding places. They are arboreal and spend the day in natural nooks and holes in tree trunks. The burrow itself is lined with a tubular silk web, the opening sealed with compact silk hairs containing antibiotics. From these a signal thread leads normally to a spider's foot. When the silk web moves, the spider is informed. As soon as you touch the cave entrance with a blade of grass or a thin stick, for example, the tarantula comes out and walks carefully, for example, on the broadside of the machete held out. This way, if you don't make any frantic movements, you can look at them in peace. By the way, their venomous bite would not be life-threatening for humans, but infections are possible due to bacteria on their biting claws.

Both sexes can be distinguished externally: the female is much larger than the male, with a body length of up to 10 cm and a wingspan of up to 20 cm. In addition, the ends of the feet, with which they taste and feel, are pinkish in the female and reddish in the male (Fig. 9.14). Incidentally, fossil specimens have been found with a wingspan of 1.80 m.

The female has a life expectancy of 12 years, but the male dies already after the last mature moult, often already during copulation.

The body and the eight legs are densely hairy. On the front part of the body, besides the legs, there are two biting or poisonous claws *(chelicerae)* and two jaw palps *(pedipalps)* as well as the small, slit-shaped mouth opening and the eye mound with eight eyes. The male has a hook on the first pair of legs to block the venomous claws of the female during mating. On the abdomen are two pairs of spin glands and barbed stinging hairs, which can be frictionally ejected and hurled at the enemy in case of danger. The spiders grow by moulting.

Tarantulas are nocturnal and sit at dusk in front of their cave and wait for prey or they undertake their forays into the surrounding area. They catch for example frogs, lizards, cockroaches or scorpions. The captured prey is injected with poison and digestive juice via the biting claws. The liquefied prey is then taken up by the sucking stomach via the mouth opening. Even skin and bones

Fig. 9.14 The female tarantula has pink foot ends. Newly hatched tarantulas still carry a supply of yolk with them

of frogs become mushy due to the aggressive digestive juice, the remains are spat out as spit balls.

To reproduce, the male cautiously approaches the larger female, drumming with his palps. When the female responds, she is ready to mate. The male jumps onto the female's back and inserts his palp ends into her sexual orifice, where he releases the previously ingested sperm into a storage pocket (spermatheca) of the female. In this act, out of ten males, eight are eaten. But even the successful ones die a natural death after the mating season. Next, up to ten eggs slip past the spermathecae, where they are fertilized. Finally, the eggs are laid on the outside and spun into a silk web (cocoon). The cocoon is well camouflaged, such as under the bark of a tree trunk and attached above the high-water line. In addition, the female guards the clutch. Once, while driving through a floodplain forest, I used a machete to remove a loose piece of bark from a tree trunk to see if anything interesting was hiding there. Under the bark was a tarantula cocoon that fell into the water. I picked it up with the landing net and took it home. The next day ten small, fully formed and viable tarantula nymphs had hatched in the glass jar (Fig. 9.14). The supply of yolk on their backs provided them with food for the next few days. Another time, while hiking in the woods, I inspected a *bromeliad* that had been flung to the forest floor with its support tree during a thunderstorm. Thereby I discovered in the funnel-shaped rosette about ten fully developed small tarantulas, each with a body size of a one-cent coin.

Worldwide, there are over 140 species of tarantulas in the tropics with the exception of Australia.

The German name "vogelspinne" (="bird spider") goes back to the german naturalist *Maria Sibylla Merian*, who painted a tarantula sitting on a dead hummingbird in Surinam in 1694. This inspired *Carl von Linné* in 1758 to name it *Avicularia*, which means "similar to a small bird". The spiders are called in English, Spain and Portuguese tarantulas, after the Spanish city of *Taranto*, famous for its frenzied *tarantella* dances.

9.10 Ants

Insects (from the Latin word *insectus* (=incised; referring to the two body parts set apart from each other) are omnipresent in the rainforest, but hardly visible—with the exception of ants. There are about 13,000 species worldwide, which is considerably more than bird species (35 ant species live in Germany). They have evolved from wasp-like ancestors and have been around for 100 million years. During this time, many different forms have evolved. Depending on the habitat, they are omnivorous, carnivorous or herbivorous. As later in human evolution, different life forms have evolved in ants. There are

- Migratory herders who carry aphids ("greenflies") or scale insects as livestock on their migrations as nomads and milk them as needed.
- Then there are ant species that have evolved into sedentary plant breeders and gardeners, growing mushrooms and living off their products.
- Finally, there are even slavers who subjugate other ant colonies by stealing their brood, scenting them with their own scents, and thus enslaving them.

In the Brazilian rainforest, the biomass of all ants is greater than that of all land vertebrates. Extrapolated, eight million ants live on one hectare (that is about 2.5 acres) of rainforest, which is 8000 kg of biomass, about as much as eight VW beetles weigh! Using the method of US biologist *Terry Erwins*, whereby the crowns of individual trees are sprayed with biodegradable *pyrethrum* so that all insects are stunned and fall to the ground, it has been determined that 46 species of ants live on one giant tree (for example, the kapok tree)—as many as there are in Great Britain in total.

They are indeed the secret rulers of the rainforest. For example, they eat one-sixth of all the leaves in Amazonia, thus forming 1 cm of humus soil in 100 years. And they consume more meat food than all predators combined. Without ants, the rainforest would collapse; there simply wouldn't be enough nutrients available.

The head or state mother of a colony is the queen, who lives up to 6 years and lays up to 140 million eggs in her lifetime. Every few years she produces winged males and winged females. Only 1% of these "princesses", which are fertilized by the winged males, survive and establish a new state as queen each time. Through scents imparted to all members of the state and through hormonal control, each member altruistically has a predetermined role to fulfill according to the motto: "One for all, all for one." They communicate via chemical messengers that correspond to spoken words. With their jaws, the mandibles, they are very defensive. They can use them to stab their enemies, cut off segments, or even inject venom.

Tip

To learn more about the life of ants, you can buy a starter set with a queen and some "workers" or even a whole colony in a plastic nest in an ant shop (for example *Ant Store* in Berlin). The artificial ant nests are transparent, which doesn't bother the ants, so you can watch them doing what they do. They are even ideal pets, because they neither smell unpleasant nor make noise. They always keep their burrow clean and require little space. You only have to feed them with sugar solution. Since ants do not like water, their burrows are best surrounded by a water channel, so they cannot escape.

9.10.1 Leafcutter Ants

The leafcutter ants, for example of the genus *Atta*, are of great importance for the survival of the rainforest. Over the years, their queen builds up nests with up to six million inhabitants. Since these ants cannot eat plant food directly, they have developed a method of obtaining plant nutrients in a roundabout way: they live in a close *symbiosis* with certain fungi. The ants feed the fungi, which in turn produce nutrients that the ants feast on—an ingenious way of life.

For fungi feeding, the ants have developed a perfect division of labour with different steps. First, suitable pieces of leaves of selected trees must be cut. For this purpose, 2 mm scouts, leaf cutters and transport workers are necessary on their well-groomed roads. In order to cut 1 m^2 of leaves, the workers have to cover a distance of up to 3 km—to the food source and back to the burrow. The leaf pieces are carried into the underground burrow like a green sail (Fig. 9.15). Each day, as much food is carried into the burrow as a full-grown cow eats—this adds up to about 35 t per year!

Once I observed that they carried only coloured pieces of petals to their burrow, like a variegated moving garland. Thus a never-ending stream of workers is constantly moving from their burrow to the harvesting sites, and vice versa an equally large column with the leaf pieces to the burrow. The whole thing resembles a busy highway, but there are no traffic jams with these ants.

In a small experiment, I once laid a large leaf across their clean road. Sure enough, for exactly 3 min there was a traffic jam in both directions, numerous ants running criss-cross. Then suddenly the waiting ants formed up and all the traffic moved in both directions via the diversion around the leaf back to the road and the traffic jam was eliminated.

On this road are also gigantic, 10 mm guardians, true fighting machines with enormous pincers that cannot feed themselves. In the burrow there are leaf shredders, which crush pieces of leaves to 1 mm in hundreds of chambers,

Fig. 9.15 Leafcutter ants cut off pieces of leaves with which they feed the fungi grown in their burrow. This is a roundabout way for them to obtain a diet rich in protein and vitamins. Clearly visible in the upper picture are the entrances to the underground burrows of the leafcutter ants

as well as leaf chewers, leaf stackers, cleaners and inoculators, which treat the plant mass against introduced bacteria and fungi. The shredded plant mass serves as food for special fungi that are bred in these chambers. These fungi overgrow the leaf mass with their fungal filaments *(hyphae)*. Ant harvesters repeatedly bite off these fungal threads, resulting in the formation of protein-rich, vitamin-rich, tuberous thickenings. These are eventually bitten off by harvesters and distributed to members of the state. In rearing chambers, baby-sitters and nursery workers care for the eggs and larvae. The egg-laying queen inhabits her own birthing chamber. In her more than 10 years of life, she can lay more than 100 million eggs.

The subterranean burrow of the leafcutter ants extends up to 8 m into the depth in the middle of the jungle and consists of thousands of chambers and passages with a volume of about 50 m^3. Up to 40 tons of primeval forest soil are moved in the process. From above, the structure looks like ploughed-up earth on an area of about 6 by 8 m, with a few large holes marking the entrances and exits (Fig. 9.15).

The fungal gardens are also home to large colonies of bacteria that fix nitrogen from the air. When the nests of the leafcutter ants re-enter the jungle cycle with the death of their queen, the forest obtains a considerable fertilization and thus revitalization.

9.10.2 Azteca Ants

The *Azteca* ants belong to a genus that also lives in close symbiosis, with *cecropia* trees (Sect. 6.1.1). The tree provides food and living space for the ants, and the ants in turn protect the tree from herbivorous insects and the growth of lianas, mosses and epiphytes (Sects. 6.5 and 5.10). Young *cecropia* trees, as pioneer plants, often stand for months in the water of riverbanks, lakes and várzeas (Fig. 6.1). Nevertheless, more than half of the slender saplings and somewhat older, perhaps 6 m high young trees are inhabited by colonies of these ants with millions of individuals, for example the species *Azteca muelleri.*

The close relationship between tree and ants begins when a young queen, before establishing her state, enters through the preformed thin zone of a leave scar between two nodes of the stem (internode) and begins to lay eggs there. As the number of individuals in the ant colony increases, more and more *internodes* are colonized (Fig. 9.16). Since each transverse wall of an *internode* also naturally already has a hole, the ants can easily expand their brood chambers from one *internode* to the next until the entire tree is inhabited by millions of these reddish ants barely a millimetre in size.

Fig. 9.16 Hollow *internodes* of cecropial shoots are colonized by Azteca ants

The tree offers the ants on the undersides of the leaf bases button-like, fat- and protein-rich 1–2 mm long excretions (so-called *Müller's fruiting bodies*), which are regularly harvested by the ants. The ants additionally establish scale insect colonies that feed on the tree sap. The honeydew excreted by the scale insects is eaten by the ants.

Between both species of organisms there is a co-evolution extending over a long period of time, in which one species (the *cecropia*) selects the second species (the ants) by means of changes (prefabricated holes on the shoot); this co-evolution, however, also worked the other way round (the positive influence of the ants—greater fitness of the tree—led to the production of the feeding bodies).

9.10.3 Bullet Ants

The bullet ant (also called giant ant; *Paraponera clavata*) lives up to its name and is of spectacular size: the workers grow—like their queen—to 3 cm, they are black, body and legs are hairy, the venomous sting is located on the abdomen. Their long, narrow waist between the abdomen and the forebody is striking. It resembles a wingless wasp in appearance (Fig. 9.17).

The entire state, unlike the aforementioned relatives, includes only a few dozen members. Bullet ants inhabit the trunk base of trees in the deep rainforest, where the underground nest is made. Slight, fresh earth elevations around a trunk base indicate their underground burrows. If you tap or beat with a

Fig. 9.17 The 3 cm bullet ants are very defensible with their poisonous spines

machete there or on the trunk, some workers will come out of their burrow. You can then easily pick them up with the machete and observe them, because they are not aggressive—most of the time anyway. They guard "their" tree, climbing up to its crown to feed on young leaves and flower nectar. A Brazilian biologist told me that he had climbed up a tree to pick off an epiphytic orchid. In the process, a bullet ant gave him a venomous stitch. He described the ensuing pain as if he were burning alive. Fever and cardiac arrhythmia were added. Immediate treatment with ice water relieved the greatest pain, but who has ice water in the middle of the jungle? The extreme pain lasts up to 24 h, which is why these ants are also called 24-h ants. The venom *poneratoxin* is a slow acting neurotoxin, which is particularly painful but does not cause tissue damage. On the sting-pain-index with a scale of 1–4, which measures the subjective sensation of pain, this venomous stitch is rated 4+. In the Amazon lowlands, these ants are also known as *tucandera*, which means "one who is severely wounded" in the *Tupi* language. The species name *clavata*, on the other hand, translates to "club-shaped" and refers to the shape of the ant's body.

9.11 Termites

During a hike through the rainforest, any visitor will notice the black, spherical structures, ranging in size from 50 cm to over a meter, that sit on a branch in some treetops (Fig. 9.18). They are the nests of termites, which look like small, white ants, but are not.

Fig. 9.18 Termite nests are always built in the treetops in the rainforest The termite queen has a shapeless swollen abdomen due to the huge number of constantly newly formed eggs

Their relationship to ants is as distant as that of monkeys to kangaroos. In fact, termites are related to cockroaches. There are about 2000 species of them worldwide, all at home in the tropics. Together with ants, they account for 25% of the total biomass in the Amazon lowlands. They play an important role in the rainforest. The name "termite" is derived from the Greek word *termes*, which means "end" and refers to the wood-destroying activity of these animals, as they can "put an end" to any wood: they eat—mostly dead—wood. However, they cannot digest the chewed wood fibres themselves. For this they depend on the help of fungi and bacteria that live in their rectum and process the cellulose for the termites so that they can live on it. This method of nutrition returns many of the wood's constituents to the material cycle of the rainforest, and termites also fertilize the forest with their excrement. You could also say that they recycle dead wood and, together with the ants, take over the role of earthworms, which do not exist in the sodden jungle soils.

Because of the frequent floods, their spherical nests hang exclusively in the trees and are not built on the ground as, for example, in the dry regions of Africa. In Amazonia, these structures are built of chewed wood, inside there are corridors, living and rearing chambers throughout the entire structure, the structure is constantly ventilated and functions like an air conditioner. The air that heats up in the burrow constantly escapes upwards, so that cooler air can flow in from below through small openings. Walkways covered with chewed wood lead from the burrow directly to the food sources. If you open such a passage with the fingernail, immediately numerous termites appear to close the hole. The "feeding trees" are literally hollowed out from the inside by eating the wood; the tree cover remains for a longer time.

The termite state is founded by a couple, the inconspicuous king and the queen. Due to the enormous number of constantly newly formed eggs, the abdomen of the queen swells shapelessly to several centimeters in length (Fig. 9.18).

It then resembles an "egg-laying machine" that lays an egg every 8–10 s, that is up to 10,000 a day and 300,000 a month. In its life, which can last up to 10 years, it lays many millions of eggs in her "nuptial chamber"! These must be fertilized again and again by the king. These eggs develop—controlled by hormones—into various termite castes, which are differently designed groups with different tasks:

- The blind, 2 mm large workers are responsible for brood care, building the nest, providing food and looking after the queen.
- The soldiers, also blind and measuring up to 5 mm, defend the workers as they forage and fend off intruders.
- The winged sexuals, up to 2 cm in size, develop only about every 2 years. They then swarm out, where they are threatened by many predators. Fertilization takes place in the air, from which, with luck, a new pair of kings emerges.

9.12 Beetles and Praying Mantises

The giant ***actaeon* beetle** *(Megasoma actaeon)* belongs to the rhinoceros beetles. With a body length of 15 cm and a weight of 220 g in males, it is one of the largest beetles ever. Its wingspan is 30 cm (Fig. 9.19)! On the anterior segment of the thorax, a defensible horn, about 4 cm long, projects forward; to the left and right of it sit two more short horns. The female remains clearly smaller with about 4 cm and has no horns.

Actaeon beetles are very defensible and have barbed spurs and spines on their legs, the last leg sections have claws. Their body colouration is black-brown. For reproduction, the females lay their eggs in the substrate of rotten trees, the development of the larvae lasts almost 3 years, then they weigh 200

Fig. 9.19 The wingspan of the male giant rhinoceros beetle is 30 cm

g. After metamorphosis, the adult beetles live only 5 months; during this time they feed on roots and flowers.

During my travels I have encountered these beetles a few times. Once a male landed in the dark—they are nocturnal—in the middle of the Amazon river on deck of a ship. I have also seen them in the hands of children who use them as "toys" in the villages. This involves cutting off the defensible horn and walking the beetle around the village on a leash. Finally, I spotted another male specimen in *Iquitos* on a tree on the waterfront.

9.12.1 Praying Mantises (*Mantidae)*

In the Amazon lowlands, the visitor's eye falls every now and then on a praying mantis, no matter how well it may be camouflaged in its light green ground colouration. Also, village children often bring them by or these animals suddenly sit in the boat after it has passed through floodplain meadows.

The female is with about 7 cm larger than the male and of mostly light green colouring. The triangular, strongly mobile head with two hair-shaped

Fig. 9.20 The praying mantis is a lurking hunter that waits until a suitable victim passes by

antennae and small black eyes is striking. The strong abdomen sits on transparent wings. The four hind legs serve as striding legs, the two front legs are transformed into tentacles with many spines and a spiny claw on each leg. The lower legs can be folded against the thigh like a pocket knife. This posture is also their normal resting position among green stalks (Fig. 9.20).

The praying mantis is a lurking hunter: it waits until a suitable victim (insects or spiders) passes by, and then strikes from ambush with its claws.

Mantises are also famous for their sometimes practiced sexual cannibalism, as the smaller males are often eaten during or after copulation. The 100–200 eggs are laid on a shoot in a hardening mass of foam. *Mantises* live for only 1 year.

10

Concluding Note: What Is the Future of the Amazon Lowlands?

Nothing is more difficult to predict than the future. The consequences of human actions and global events in the Amazon region that will occur in the foreseeable future can only be estimated from data from the recent and longer past. The facts available so far lead to the following prognosis: the prospects for preserving the unique ecosystem of the Amazon lowlands are not good.

The diversity of plant and animal species built up over millions of years is acutely threatened. On the one hand, the already destroyed area are immense. On the other hand, the destruction of the forest continues. In addition, two further potential threats to the Amazon basin continue to build up until one day the development will be irreversible: climate change and population growth.

10.1 Forest Destruction

The worldwide destruction of the primary forest continues unabated in Amazonia and apparently cannot be stopped by anyone. At the UN climate summits in Rio (1992) and New York (2014), forest declarations were adopted that explicitly included the Amazon rainforests. But these declarations have achieved nothing. According to Greenpeace estimates, by 2008 more than 20% of the original 3.5 to 4 million square kilometres of primary forest in the Amazon lowlands had been destroyed. After that, an area the size of Belgium (30,688 km^2) continued to be destroyed year after year and still is today. This means that within 12 years an area of primeval forest the size of Germany respectively of the US-State of Arizona is lost. If this level of forest destruction

L. Staeck, *Fascination Amazon River*, https://doi.org/10.1007/978-3-662-64452-2_10

continues, the Amazonian rainforest area of about 2.7 million km^2 that currently exists (2019) will no longer exist in 2100. The Brazilian government is doing far too little to prevent this *worst-case scenario.* In the past, it has passed nature conservation programmes from time to time in order to formally stop the overexploitation of the Amazon rainforest, for example in 1988 under the title "Our Nature". According to this, subsidy payments for large livestock farms in the Amazon region were to be stopped and a ban imposed on the export of untreated tropical timber. But it did nothing to stop the ongoing large-scale deforestation as the figures show. In the recent past, Brazilian governments have even relaxed the stricter forest laws again, so that in addition to illegal deforestation and the equally illegal burning of forests, legal logging is now being stepped up again. The powerful lobbyists in the Brazilian parliament, the lack of control of existing laws and the massive corruption in the country will encourage the further destruction of the forests. As a result, more and more animal species will disappear, hitherto existing food chains will collapse, and the remaining animals will no longer be able to find food or have any means of retreat. As shown in this volume, a primary forest cannot be "replanted", and the intact primeval forest is more than the sum of its inhabitants, for many more members of the forest are interrelated and interdependent than was suspected decades ago. Thus, the destruction of the forest will also result in the disappearance of more and more fish species that are indispensable for the population. Even now, important food fish of the Amazon region are considered overfished and their existence is threatened. It is not yet possible to predict what this will mean for the growing population in the Amazon basin. With the loss of the primary forest, however, the primeval forest soils with their *mycorrhizae* are also irretrievably destroyed, leaving behind a barren, depleted land with a red laterite cover. This problematic situation was also referred to in this volume.

10.2 Climate Change

Finally, there will also be a drastic reduction in rainfall—with a slight time lag—with dramatic consequences for the forests that still exist, as well as for the animals that still live in these forests and also for the fish fauna. This prediction is derived from the fact that normally 50% of the rainfall that falls on a rainforest is regenerated from the evaporation of the rainwater itself. However, when the forest is no longer present, this rainfall quantum ceases. The other 50% of the rain so far is fed by the trade winds from the Atlantic. With the progressive loss of forests, this rainfall now reaches the rivers flowing off to the Atlantic at an ever-faster rate, since the closed canopy of the forest, which previously slowed down the rain, is no longer present and the sponge effect of the primeval forest soil matted with roots and mycorrhiza can no longer take effect.

These considerations do not even take into account the possible effects of the global climate change that is already taking place, as they are not even foreseeable at present. No one knows how *El Niño, La Niña* and other climate phenomena like the Pacific Decadal Oscillation (PDO) and the solar cycle of sunspot activities will affect the constantly shrinking forest area of the Amazon lowlands—and vice versa. One thing is certain, however: we are in a geo-historical warm phase after the last ice-age on one side and the "Little Ice Age" (LIA) from the sixteenth to the nineteenth centuries on the other side with an uncertain outcome. In addition, we humans are fueling this phase with ever-increasing emissions of CO_2 and methane into the atmosphere. The entire development seems to be leading to a self-accelerating process, for which *albedo effects* (ice-albedo feedback) and the advancing warming of the oceans are largely responsible.

10.3 Population Growth

Another acute threat to the Amazon rainforest is the rapidly increasing population of the Amazon lowlands. Meanwhile 20 million people live in this metropolitan area. This population increase is fed on the one hand by the rural exodus from the dry northeast of Brazil and on the other hand by the high birth rate of 0.75% per year of the local population. Thus, the number of inhabitants in the Amazon lowlands increases by more than 350,000 year after year. On my travels along the Amazon, I witness from one year to the next how the village communities and also the population in the small towns along the river increase greatly. During my visits to schools, teachers tell me that more than half of the villagers are children and teenagers of school age. These adolescents will marry within the next 10 years and then will also have offspring. Families with more than four children are still normal among the *Caboclos* as well as among the indigenous groups, although puerperal and infant mortality rates have declined sharply and overall medical care and thus life expectancy have increased greatly.

This increase of population increase is like a ticking "time bomb". No one can estimate how long the sensitive ecosystems of the Amazon will be able to withstand these enormous human growth rates without collapsing. The big problem of exponentially growing systems such as humans can be seen in the typical course of an exponential curve: it initially rises only slowly for a long time. Even with the growth of mankind, a "deceptive calm" prevailed for a very long time, mankind and the consumption of resources grew in a reasonably manageable way, most of the system components—for example, the inhabitants of the Amazon region, the stock of fish in the rivers, the available wild and grazing animals—functioned without disruptions. However, the exponential curve then suddenly—without warning—steeply rises until it unexpectedly and suddenly tips over. It is exactly the same with us, humans, at the moment: The growth rates are getting bigger and bigger in shorter and shorter periods of time. No one knows when the point will be reached at which the system, the global ecosystem *Earth*, will topple over, collapse. One conclusion is certain: it will not go on like this forever. The following situation is currently clearly evident in the Amazon region:

The threatening increase of the population in the Amazon rainforest ecosystem results in a massive loss to its resources as a result of

- of forest destruction for the establishment of plantations, pastures, villages, infrastructure;
- the exploitation of forests through unsustainable hunting of fish, birds, mammals and uncontrolled logging;
- the destruction of the environment through water pollution, soil destruction, waste accumulation

However, the exponential population increase is not only regional but also global. The resulting negative impacts ultimately affect all people, including those in the Amazon, as they undo all efforts to reduce CO_2-emissions, preserve protected areas or conserve resources.

The world's population was about 7.9 billion at the end of 2021, up from just three billion in 1960. Annually, the world's population is currently increasing by about 80 million per year, 220,000 people every day, one million every four and a half days, and 2.5 people every second discounting deaths. The United Nations estimates that by 2050 our planet will have the following total population:

- up to 11 billion (assuming an unchecked average number of children per woman of currently 2.56 children);
- 9.1 billion (assuming an average number of children per woman of 2);
- 8 billion (assuming an average number of children per woman of 1.5);
- 6.5 billion (assuming an average number of children per woman of 1).

By the way: According to a forecast by the United Nations, the world population would be 300 billion in 2150 if the current birth rate of 2.56 per woman remained unchanged for the next 150 years.

How the number of children per family could be reduced, however, is still completely open. However, a discourse on population policy must begin as soon as possible. So far, however, this problem area is still a taboo subject. So far, neither politicians or scientist nor journalists thematise or discuss the unchecked growth of the world's population. However, one point is already certain: a further exponential increase in the world's population can only be stopped by improving the quality of life in poor societies and by fundamentally increasing the education of women in these societies.

If the fascination of the Amazon rainforest is to be preserved, then the existing positive political, economic and socio-political approaches must be bundled and the global efforts of nature conservation must be intensified in order to work through the identified problems. Above all, the global community must provide the Brazilian government with ways and great financial resources to conserve this unique Amazonian biodiversity. The preservation of this biological diversity is not an end in itself but makes a significant contribution to stabilising the economic, social and ultimately also security situation in numerous countries.

The future can be wonderful—or it can not happen at all. That depends entirely on us.

Index

L. Staeck, *Fascination Amazon River*, https://doi.org/10.1007/978-3-662-64452-2

C

MIX
Papier aus verantwortungsvollen Quellen
Paper from responsible sources
FSC® C105338

FSC
www.fsc.org

If you have any concerns about our products,
you can contact us on
ProductSafety@springernature.com

In case Publisher is established outside the EU,
the EU authorized representative is:
Springer Nature Customer Service Center GmbH
Europaplatz 3, 69115 Heidelberg, Germany

Printed by Libri Plureos GmbH
in Hamburg, Germany